Designing Support

FROM HELPER TO INSTRUCTIONAL PARTNER

LASHAE LEE, PHD
ILLUSTRATIONS BY CHRISTINE LEE

HUMAN LEARNING BY DESIGN™

Human Learning by Design™

A series exploring learning, leadership, and human development
in an age shaped by artificial intelligence.

AI and the Art of Teaching
AI and the Art of Teaching Workbook
Designing Schools
Designing Support

Forthcoming titles in the Human Learning by Design™ series include:
Designing Confidence
Designing Inquiry
Designing Judgment
Designing Learning Environments
Designing Professional Identity

The developmental Human Learning by Design™ series include:
Becoming Human: Early Childhood
Becoming Human: Middle Childhood
Becoming Human: Adolescence
Becoming Human: Early Adulthood
Becoming Human: Adulthood
Becoming Human: Later Adulthood

Related titles in the Human Learning by Design™ collection:
Raising Humans by Design
Courage Under Algorithm for Teens
Judgment Over Speed: Leadership in an AI Workplace

ISBN: 978-1-972397-06-0 (paperback)
ISBN: 978-1-972397-07-7 (e-book)

Illustrations and cover design by Christine Lee

First Edition, 2026

This book is intended for educational and informational purposes only. The views expressed are those of the author and do not necessarily reflect the policies or positions of any affiliated institutions.

Published and printed in the United States of America

CONTENTS

Designing Support

DEDICATION

For the adults beside the desk. For the paraeducators, instructional aides, and classroom assistants who notice the hesitation before anyone else does. When this book says paraprofessionals, it means you – regardless of your job title.

You kneel beside the student who is stuck. You offer quiet encouragement when frustration rises. You redirect gently so dignity stays intact. You remain steady when the work becomes difficult.

Much of what you do happens out of the spotlight. But it matters more than most people ever see.

In *AI and the Art of Teaching*, I share a moment from my early years in the classroom when I was struggling as a new teacher. The paraeducator working alongside me offered steady, thoughtful support. She helped me slow down, reflect, and find my way back to what mattered most.

That moment stayed with me. It reminded me that the most powerful support is not the answer – it is the question that helps someone think again.

You are not "just support." You are builders of confidence, protectors of dignity, and quiet architects of independence.

This book is for you.

PROLOGUE

The Adult Beside the Desk

The room is quiet in the way classrooms become quiet during independent work. Pencils move slowly across paper. Pages rustle. A few students stare at their desks, thinking. Others stare at the ceiling, hoping an idea might appear there instead.

At one table, a student sits hunched over a worksheet. The eraser has already left faint gray smudges where answers once were. An adult sits beside the desk. The student glances sideways, then leans closer so no one else can hear. "Can you just tell me the answer?" It is not asked with defiance. It is not asked with laziness. It is asked with quiet hope.

For the adult beside the desk – the paraeducator, instructional aide, or classroom assistant – this moment is familiar. It happens dozens of times each day. Often quietly, almost invisibly, in the spaces between the teacher's instructions and the student's work.

But inside this moment lives a decision. Should you rescue the student so they can move forward? Or should you slow the moment down so they can think? Should you protect dignity by helping quickly? Or build stamina by allowing a little struggle? Should you give help? Or cultivate independence?

There is rarely time to analyze the decision. The student is waiting. The pencil is paused. Silence stretches just long enough to feel uncomfortable.

Many support professionals were trained to help students succeed, and helping often feels like giving the right answer, correcting mistakes quickly, or guiding the student step by step until the problem is finished. The worksheet gets completed. The task moves forward. The moment passes.

But learning does not always grow in the same direction as efficiency. Students do not become confident because someone provides answers. They become confident when they experience themselves thinking. They become confident when they attempt something difficult and survive the uncertainty. When they revise an idea. When they discover they were capable all along.

Those moments are fragile. They can be strengthened – or unintentionally erased – by the adult sitting beside the desk.

This matters even more now than it did a decade ago. Today's students live in a world where answers appear instantly. Artificial intelligence can generate explanations, solve problems, and produce polished writing in seconds. Information is abundant. Fluency is easy to simulate. Correct answers are no longer difficult to find. But something important remains rare. Confidence. Endurance. Ownership of thinking.

Technology can generate answers. It cannot build identity. It cannot notice the hesitation in a student's voice. It cannot sense the difference between frustration that builds growth and frustration that shuts a student down. And it cannot quietly protect a student's dignity while helping them try again.

That work belongs to the human being sitting beside the desk.

Paraeducators often underestimate the influence they hold. Because the teacher leads the lesson, it can feel as though the real instruction happens at the front of the room. But the moments beside the desk are where many students decide who they are as learners.

When you kneel beside a student and say, "Show me what you've tried," you are teaching something. When you pause instead of answering immediately, you are teaching something. When you allow a student to wrestle with an idea before stepping in, you are teaching something. And when you communicate, quietly but clearly, "I believe you can do this," students begin to believe it too.

This book was written for the adults who sit beside the desk. For the paraeducators who notice confusion before it becomes frustration. For the instructional aides who redirect behavior quietly so learning can continue. For the support professionals who work closest to students who often feel least confident.

Too often, the work of paraeducators is described as helping.

But helping is not the deepest purpose of the role. You are not an answer provider.

You are a confidence architect. You are a protector of thinking. You are part of the human layer of the classroom – the part no technology can replace.

This book will help you approach that work with clarity and confidence.

You will learn how to:

- Support students without replacing their thinking
- Build independence instead of dependence
- Interpret behavior wisely
- Protect dignity during struggle
- Work in alignment with the teacher's instructional design
- Use simple questions that strengthen reasoning and confidence

You do not need to design the lesson to shape the learning.

Sometimes the most powerful teaching moment in a classroom happens quietly – between a student, a difficult problem, and the adult sitting beside them.

And when the student whispers, "Can you just tell me the answer?" you will know how to respond. Not because you refuse to help. But because you understand something deeper.

Your job is not to give the answer. Your job is to help the student discover they can find it.

WORKBOOK PHILOSOPHY

This Book Is About Forming Thinkers.

Many professional development books promise a collection of strategies, quick fixes, or classroom tools. This book is different.

It is not primarily about adding more techniques to your day. Most paraeducators already carry a long list of responsibilities: redirecting behavior, clarifying instructions, helping students organize materials, supporting reading or writing, and keeping lessons moving when frustration appears. Adding more tools will not necessarily make that work easier. And still, there are many tools in this book for you to consider. This book focuses on something deeper: how students experience themselves as thinkers when you are beside them.

Every interaction with a student sends a message. When you lean down beside a desk and quietly offer help, the student is not only hearing your words. They are also interpreting what your help means about them.

They may be learning: "I can't do this without help." "Someone will fix it for me." "Struggle means I'm failing." Or they may be learning something very different: "I can try." "I can think." "I can figure things out." "Struggle is part of learning." The difference rarely comes from a complicated strategy. It usually comes from how help is offered.

This book is built on a simple idea: Support should strengthen thinking, not replace it. That means the goal is not simply helping students complete assignments. The goal is helping students develop the confidence and stamina to think through challenges themselves.

Sometimes that looks like asking a question instead of giving an answer. Sometimes it means pausing instead of stepping in immediately. Sometimes it means stepping away for a moment so a student can attempt the work independently.

Those choices can feel uncomfortable at first. Many support professionals care deeply about students and want to prevent frustration or failure. But growth rarely happens when thinking is removed from the process.

Students become more capable when they experience themselves solving problems, revising ideas, and trying again. Your role helps create those experiences.

This workbook is designed to help you notice the moments where thinking either grows or disappears. Each chapter focuses on a common situation you encounter every day: A student asking for the answer. A student refusing to try. A student becoming frustrated during difficult work. A student depending on adult help before attempting independently

Instead of providing complicated programs, the book offers something more practical: Clear ways to interpret what is happening, and simple ways to respond that build independence and confidence over time.

The goal is not perfection. The goal is intentional support.

You will not need new equipment, new software, or complex lesson plans to apply what you learn here. Most of the changes involve small shifts in how you interact with students during the moments you already experience every day. Small shifts can produce powerful results.

When support strengthens thinking instead of replacing it, students begin to experience themselves differently. They attempt more often. They persist longer. They rely less on immediate help. They begin to see themselves as capable learners. That is the deeper purpose of this work.

In classrooms shaped by powerful technology and increasing academic demands, answers are easier to generate than ever before. What remains difficult – and deeply human – is developing the confidence to think independently.

Paraeducators play a critical role in that process.

PART I – YOU ARE MORE THAN SUPPORT

Chapter 1: The Hidden Curriculum of Proximity

Many people assume that the most important teaching happens at the front of the classroom. That is where the lesson is introduced. That is where directions are explained. That is where the content appears on the board.

But much of what students learn about themselves as learners happens somewhere else. It happens beside the desk.

Paraeducators often work closer to students than anyone else in the room. You sit beside them when the assignment becomes confusing. You hear the quiet questions students are sometimes too embarrassed to ask out loud. You notice frustration long before it turns into shutdown or behavior.

Your proximity gives you a unique influence over how students experience learning. You are closer to some students than anyone else in the room. You kneel beside their desk. You whisper clarifications. You redirect behavior quietly. You restate instructions. You help organize papers. You notice frustration before it erupts.

And in every one of those moments, you are teaching. Not just content. Identity.

Students are constantly interpreting what adult support means about them. If an adult always steps in quickly, a student may begin to believe they cannot work independently. If answers appear before effort begins, a student may assume thinking is optional. If help always arrives immediately, struggle may begin to feel like failure instead of part of learning.

But proximity can communicate something far more powerful.

When support encourages students to attempt first, to explain their thinking, and to stay with a challenge a little longer, students begin to experience themselves differently. They start to believe they are capable of figuring things out.

That belief grows slowly, one interaction at a time. This is what we call the hidden curriculum of proximity.

Students are not only learning the academic task in front of them. They are learning what adult support means about their capability, independence, and place in the classroom. The adult beside the desk helps shape that message.

And when proximity is used intentionally, it becomes one of the most powerful tools for building confidence, protecting dignity, and strengthening independent thinking.

You are not just helping students complete work. You are shaping how they see themselves as learners.

Proximity Is Never Neutral

When you sit beside a student, you are not simply offering help. You are sending signals. Signals about: Intelligence. Capability. Power. Dependence. Expectation. Belonging.

Students interpret those signals quickly. Sometimes faster than adults realize.

They notice how often help arrives. They notice what kind of help they receive. They notice whether the adult expects them to try first or waits for them to ask.

Other students in the room notice these signals too.

They watch who receives support. They watch how often help appears. They watch what happens when someone struggles. Over time, these patterns begin to communicate messages about who is capable and who is not.

This is why proximity is never neutral. The same act – sitting beside a student – can communicate two very different messages.

It can unintentionally communicate: "You can't do this alone." "You need someone next to you to succeed." "The work is not really yours." "Someone else will fix it if it becomes difficult."

Or it can communicate something much more powerful: "I believe you can do this." "I will help you think, not replace your thinking." "I am here while you figure it out." "You are capable of growing stronger."

The difference is not in where you sit. It is in how you sit there.

Small decisions shape the message students receive.

- Do you provide the answer immediately, or ask what they have tried first?
- Do you stay beside the student for the entire assignment, or step away so they can attempt independently?
- Do you correct the work yourself, or ask the student to explain their thinking?

These moments may feel small, but they accumulate. Over time, they shape how students experience themselves as learners.

When proximity is used carefully, it strengthens confidence. When it is used automatically, it can unintentionally create dependence.

The goal of support is not simply to make work easier. The goal is to help students become stronger thinkers.

And the adult beside the desk plays a powerful role in making that possible.

The Signals Students Receive

Let's make this concrete.

When you sit beside a student, the help you provide sends signals about what you believe they are capable of doing. Students rarely analyze these signals consciously, but they absorb them quickly through repeated experience.

Consider what students may learn when support looks like this.

When you provide answers quickly, write responses for a student, speak for them during discussion, reduce task difficulty without teacher direction, step in before productive struggle, or constantly hover nearby.

Students may begin to internalize messages such as: "I can't do this." "I shouldn't try." "Someone else will fix it." "Struggle means failure." "Help means rescue."

None of these messages are intentional. Most adults step in quickly because they care deeply about the student and want to prevent frustration.

But over time, patterns matter.

Now consider what students may learn when support looks different.

When you ask probing questions, pause instead of filling silence, encourage a first attempt, normalize struggle, fade support intentionally, and protect dignity publicly students begin to internalize something else: "I can try." "I can think." "I can persist" "I can improve." "I am capable."

The difference between these two outcomes rarely comes from complicated strategies. It usually comes from small decisions made in everyday moments.

Proximity magnifies your influence.

You are not just helping students complete tasks. You are shaping how a student sees themselves as a learner.

The Risk of Over-Support

Paraeducators often feel pressure to "make sure the student succeeds."

That pressure can come from many places – teachers, expectations from families, school routines, or simply the desire to help a student avoid frustration.

But it raises an important question: What do we mean by success?

Is success finished work? Correct answers? Quiet compliance? Fast completion? Or is success something deeper? Is it increased independence, stronger self-belief, more cognitive stamina, or a willingness to attempt without prompting?

These outcomes do not always appear immediately. In fact, building independence can sometimes look messier at first. Students may hesitate longer, attempt imperfectly, or struggle before finding their way forward.

But that struggle is often where learning grows.

Over-support can unintentionally create learned dependence.

If a student never experiences productive struggle without immediate intervention, they never build endurance.

If a student always looks to an adult before thinking, thinking gradually weakens.

If help arrives before effort, effort begins to fade.

Support should build strength – not replace it.

The goal of support is not to remove difficulty. The goal is to help students develop the confidence to move through difficulty.

Proximity Impact Reflection

Take five minutes and reflect honestly. Think about one student you support regularly.

Then answer the following questions.

1. When I sit beside this student, what typically happens?
2. Do I speak more than they do?
3. Do I intervene before confusion deepens?
4. Does the student look at me before attempting the work?
5. Do they attempt independently when I step away?
6. If I were absent for a week, would their independence increase, decrease, or remain stable?

Now ask the hardest question: What is my proximity currently teaching this student about themselves?

Write your answer without defensiveness. Clarity precedes growth.

Recognizing patterns is not about guilt or blame. It is about becoming more intentional with the influence you already have.

Because the adult beside the desk is never neutral.

Your presence can quietly shape confidence, independence, and identity – one interaction at a time.

Independence vs. Over-Support Diagnostic

Support professionals often step in quickly because they care about the student and want the work to go well. That instinct is understandable. But over time, patterns of help can either strengthen independence or quietly replace it.

This quick diagnostic helps you notice which pattern appears more often in your current practice. Select the statements that reflect what you typically do first when a student begins to struggle.

I tend to:

☐ Restate directions immediately
☐ Provide sentence starters before the student attempts an answer
☐ Correct errors as they occur
☐ Reduce task complexity quickly
☐ Give hints within five seconds of confusion
☐ Stay physically beside the student for most of the lesson
☐ Answer questions that could be redirected back to the student

Now look at the second list and select the statements that also reflect your practice.

I also tend to:

☐ Allow silent think time before intervening
☐ Ask "What do you think?" before offering help
☐ Step away intentionally so the student can attempt independently
☐ Return later rather than remaining beside the student
☐ Praise effort or strategy before focusing on correctness
☐ Encourage peer assistance when appropriate
☐ Allow mild confusion to exist briefly before stepping in

Now compare the two lists. Which column has more items selected?

If the first column contains most of your circles, your proximity may unintentionally be creating dependence.

If the second column contains more circles, your proximity is likely building cognitive stamina and independence.

Neither list is meant to produce guilt or self-criticism.

Every support professional moves between both patterns at times. Students need guidance, reassurance, and clarification throughout the day. The goal is not to eliminate help.

The goal is to align help with independence.

Sometimes the most supportive action is not answering immediately. It is giving the student a moment to think, attempt, and discover what they can do on their own.

Support that strengthens thinking builds confidence.

And confidence is what allows students to keep trying long after the assignment is finished.

"What Does My Help Communicate?" Checklist

Before stepping in to help, pause for a moment and consider the message your support might be sending. Support is not only about solving the immediate problem in front of the student. It is also about shaping how the student experiences themselves as a learner.

A brief pause can help ensure your help strengthens thinking rather than replacing it. Before helping, ask yourself:

1. Am I preserving the student's dignity?
2. Am I increasing independence?
3. Am I reinforcing the teacher's expectations for the lesson?
4. Am I supporting the student's thinking — not replacing it?
5. Am I stepping in because it is easier — or because it is truly necessary?
6. Am I modeling calm rather than urgency?
7. Am I helping the student maintain ownership of the work?

You will not always answer "yes" to every question. Students sometimes need direct guidance, clarification, or reassurance. But these questions help shift the goal of support from task completion to student growth.

If the answer to #2 (increasing independence) or #4 (supporting thinking) is no, it may be worth reconsidering how you help in that moment. A small change – such as asking a question first, pausing before speaking, or encouraging a first attempt – can protect the student's role as the thinker.

Help that removes thinking weakens identity. Help that strengthens thinking builds confidence. And confidence is what allows students to continue learning long after the adult beside the desk steps away.

The Hidden Curriculum Other Students See

Proximity affects more than the student beside you. When you sit next to a student regularly, the rest of the classroom notices. Students observe adult behavior closely, even when it appears they are focused on their own work.

They notice who receives help, how often help arrives, how quickly support appears, and the tone used when help is given. Over time, these patterns begin to send messages to the entire class.

If proximity becomes constant and highly visible, other students may quietly interpret it in ways that were never intended. They may think: "They can't do it." "They get special treatment." "They don't have to try as hard."

These interpretations can affect both the supported student and their peers. A student receiving help may feel exposed or different. Other students may misunderstand the purpose of the support.

Your job is not to erase support or hide it completely. Support is necessary, and many students benefit from additional guidance. The goal is to design support in a way that protects dignity and communicates growth rather than limitation. Sometimes this means adjusting how proximity appears in the room.

You might rotate proximity rather than staying beside one student continuously, support discreetly through brief check-ins, encourage peer partnerships when appropriate, or gradually fade physical closeness as independence grows. These small shifts help communicate a powerful message: support exists to strengthen learning, not replace it.

In classrooms, students are always interpreting what adult actions mean. Visibility matters. Tone matters. Consistency matters.

When proximity is used thoughtfully, it can build confidence without diminishing dignity – both for the student receiving support and for the classmates who are watching and learning alongside them.

From Helper to Confidence Architect

Many support professionals are described as helpers. Helping students is certainly part of the work. You clarify directions, organize materials, redirect behavior, and step in when confusion appears. But the deeper role is not simply helping students finish tasks.

You are not an answer provider. You are a confidence architect.

Confidence does not appear because a student receives the correct answer. It grows through repeated experiences where students discover they can attempt, struggle, revise, and improve. Confidence grows through attempt, struggle, feedback, revision, and ownership. These experiences form the foundation of independence.

When students attempt something difficult, they begin to see themselves as capable. When they struggle and persist, they build endurance. When they revise an idea and improve it, they develop belief in their ability to grow.

But when these experiences are removed, confidence cannot develop. If you remove attempt, you remove growth. If you remove struggle, you remove endurance. If you remove ownership, you remove agency.

AI and the Art of Teaching explains that in a world where intelligent tools can generate answers instantly, the most important work in classrooms is not producing answers – it is cultivating thinkers. Paraeducators play an important role in protecting that work.

When you ask a student what they think before offering help, you protect their role as the thinker. When you allow a student to wrestle with an idea before stepping in, you protect their opportunity to grow stronger.

Your presence should increase thinking, not reduce it. You are not simply helping students complete the work in front of them. You are helping them develop the confidence to believe they can think for themselves.

IMPLEMENTATION

Redesign One Support Routine

Change in classrooms rarely begins with large programs or complicated plans. More often, it begins with small shifts in everyday routines. Paraeducators interact with students dozens of times throughout the day. Each interaction is an opportunity to either reinforce dependence or strengthen independence. Rather than trying to change everything at once, start with one routine you use regularly.

Choose a routine you support daily. Examples might include: Independent work support. Reading assistance. Behavior redirection. Writing scaffolding. Math problem-solving support. Once you have chosen a routine, use the following steps to examine and redesign how support appears in that moment.

Step 1: Identify the Current Pattern

Think about what typically happens when a student begins to struggle. Ask yourself: What do I usually do first? Do you restate directions? Offer hints immediately? Provide the answer quickly so the student can move forward?

There is no judgment in this step. The goal is simply to notice your current pattern. Awareness is the starting point for intentional change.

Step 2: Identify the Hidden Message

Next, consider what that pattern might communicate to the student.

Ask yourself: What message might my help send about the student's capability? For example, if help arrives immediately, the message might be: "Someone else will fix this for me.". If the adult asks questions first, the message may be: "My thinking matters.". Students are constantly interpreting adult behavior, even when we are not aware of it.

Support always communicates something.

Step 3: Redesign for Ownership

Now experiment with small shifts that place the thinking back in the student's hands. Instead of taking over the work, try prompting the student to engage first. For example:

Replace "I'll help you" with "Show me what you've tried."

Replace "Here's how you do it" with "What do you think the first step might be?"

Replace immediate correction with "Tell me your reasoning."

Replace staying beside the student continuously with "I'll check back in five minutes."

These changes may feel small, but they shift the student's role from passive receiver to active thinker.

Step 4: Implement for One Week

Try this redesigned routine for one week. As you do, observe what changes. Notice: Did the student attempt the work more often? Did frustration increase briefly before improvement appeared? Did the student begin working independently for longer periods? Did their explanations or reasoning deepen? Did it feel uncomfortable for you to step back at first?

These reactions are normal. When students are used to immediate help, independence can initially feel unfamiliar. They may hesitate, look for reassurance, or need time to adjust.

That does not mean the support is failing. It often means the student is beginning to experience something new: ownership of their thinking.

Growth rarely looks smooth in the beginning. Sometimes it looks uncertain. Sometimes it looks slower. But over time, these moments build confidence, stamina, and independence. And those are the outcomes that matter most long after the assignment is finished.

Closing Reflection

You are not "just" support. You are more than the adult beside the desk. You are often the first person to notice when a student becomes unsure, frustrated, or ready to give up. You are close enough to hear the quiet questions and see the hesitation before it turns into discouragement. In those moments, what you do matters.

In an AI-shaped world where answers can be generated instantly, your role becomes even more important. Machines can provide solutions. They cannot cultivate identity. They cannot interpret hesitation. They cannot protect dignity. They cannot sense when struggle is productive versus overwhelming.

You can.

Your presence allows you to notice the difference between a student who needs reassurance and a student who simply needs a moment to think. You can encourage effort, ask a question that redirects thinking, or step back just long enough for the student to discover they are capable. Those moments build confidence.

Your proximity is powerful. Use it to strengthen thinking, not replace it. Use it to build independence, not dependence. Because what students learn about themselves when you sit beside them may shape them far longer than the worksheet ever will.

Chapter 2: Confidence is Infrastructure

Confidence is often misunderstood. Many people assume confidence is a personality trait. Something a student either has or does not have. But confidence is not personality. It is not charisma. It is not speed. It is not loud participation. Confidence is something much more practical. Confidence is patterned experience under designed conditions.

Students become confident when they repeatedly experience themselves attempting something difficult, receiving feedback, revising their thinking, and realizing they can improve. Those experiences form patterns. And those patterns shape belief.

As a paraeducator, you help create those patterns every day. You are not just supporting assignments. You are helping shape the conditions under which students experience themselves as learners. In other words, you are helping build – or weaken – confidence infrastructure.

Infrastructure is what holds when pressure increases. It is what supports a student when the work becomes difficult...when peers are watching...when answers are uncertain...when mistakes become visible. In those moments, confidence determines what happens next. Does the student lean in and attempt the work? Or do they shut down, look away, and wait for someone else to take over?

Confidence determines whether students see challenge as an invitation to think or a signal to withdraw. And confidence rarely develops by accident. It grows through carefully supported experiences where effort, thinking, and revision are protected. Paraeducators play a powerful role in creating those experiences.

The adult beside the desk helps determine whether students experience struggle as something that defeats them – or something that strengthens them. This chapter explores how confidence grows, how it can unintentionally weaken, and how the daily interactions between students and support professionals help build the infrastructure that allows students to keep trying when learning becomes difficult.

Confidence, Simplified

Psychologist Albert Bandura described a powerful idea that helps explain how confidence grows. He called it self-efficacy - a person's belief that they can succeed at a specific task. Self-efficacy is not the same as self-esteem. It is not simply feeling good about yourself. It is not the belief that "I am a good person." It is something more practical. Self-efficacy sounds like this: "I can do this." "I can try." "I can improve." "I can recover if I make a mistake."

Students who develop this belief approach challenges differently. They attempt the work more readily, persist longer when things become difficult, and recover more quickly from errors. Students who lack this belief often hesitate, withdraw, or wait for someone else to solve the problem. Self-efficacy grows through repeated experiences. Bandura identified four experiences that strengthen a student's belief in their own capability. For daily classroom practice, these can be understood simply.

1. Mastery Experiences. "I tried something difficult and eventually succeeded." This is the most powerful builder of confidence. When students experience success after effort, they begin to trust their ability to learn and improve.
2. Modeling. "I saw someone like me do it." Students gain confidence when they observe peers attempting, struggling, and eventually succeeding. Seeing others work through difficulty makes the challenge feel possible.
3. Verbal Encouragement. "Someone credible told me I could improve." When a trusted adult expresses belief in a student's ability to grow, that message can strengthen their willingness to attempt difficult work.
4. Emotional Safety. "I did not feel humiliated while trying." Students are more willing to try something difficult when mistakes do not lead to embarrassment or shame.

You influence these experiences when respond by sending a message about what that struggle means. Does it mean the student needs to be rescued, or that they are in the middle of learning? Over time, these moments shape a student's belief about what they are capable of. Confidence is not built through survivable challenge – moments where students attempt something difficult, receive support, and discover they are stronger than they expected.

Bravado vs. Avoidance

Confidence does not always look calm. Sometimes it looks confident on the surface, but underneath it is protecting something much more fragile. In classrooms, students often respond to difficult work in ways that hide uncertainty rather than reveal it. Sometimes it sounds like: "This is easy." (before attempting), joking when the work becomes difficult, distracting others, refusing to begin, or saying "I don't care."

Other times it looks quieter: Shutting down. Avoiding eye contact. Delaying the task. Perfectionism that prevents starting. These reactions are often protective behaviors. When students believe that struggle might expose them as incapable, they try to protect their identity in different ways. Two responses appear most often: Bravado and Avoidance.

Bravado sounds confident, but it is often a shield. Students may say: "I already know this." "This is stupid." "I'm not even trying." Bravado protects identity by pretending that struggle does not exist. If the student never attempts the work, then failure never becomes visible.

Avoidance looks different. Instead of loud reactions, students quietly withdraw. You might see their head down, silence, delayed work, bathroom trips, or passive compliance without thinking. Avoidance protects identity by escaping risk.

If the student never engages fully, they never have to test whether they might fail. Both responses – bravado and avoidance – are attempts to stay safe. Students are trying to protect themselves from embarrassment, judgment, or the feeling of not being capable. This is why direct confrontation often makes these behaviors stronger. When students feel exposed, their protective responses increase.

Your role is not to confront these reactions harshly. Your role is to make risk survivable. When students experience classrooms where struggle is normal, mistakes are treated respectfully, and effort is supported, the need for bravado and avoidance begins to decrease. And in those conditions, confidence has space to grow.

Making Struggle Safe

Students build confidence through repeated experiences where challenge is present but dignity is protected. Confidence grows when students experience: Attempt without humiliation. Error without identity damage. Feedback without exposure. Struggle without abandonment. These conditions allow students to take the kinds of risks that real learning requires.

When students believe that mistakes will lead to embarrassment, they become cautious. They may stop attempting difficult work, wait for someone else to answer, or withdraw from participation entirely.

If struggle leads to humiliation, confidence erodes. If errors are corrected publicly without protection, confidence shrinks. If help arrives too quickly – before effort or thinking has begun – confidence never has the opportunity to form. Students may complete the task, but they do not experience themselves as capable learners.

The goal in classrooms is not to eliminate struggle. Struggle is often where learning happens. The goal is to structure struggle so it remains survivable.

Students need to know that difficulty does not mean failure, and that mistakes are part of the learning process rather than a signal of weakness.

Paraeducators play an important role in creating these conditions. You are not there to remove difficulty. You are there to protect dignity during it.

When students know they can attempt something difficult without being embarrassed or abandoned, they begin to take the risks that lead to growth.

And in those moments, confidence begins to take root.

Strategy Naming Script

One of the most powerful ways to build confidence is to notice and name the strategies students use while learning.

Many adults praise outcomes. We say things like: "Good job." "Nice work." "You got it right."

Encouragement matters, but outcome-based praise can sometimes send the wrong message. Students may begin to believe their success depends on getting the answer correct rather than on the thinking that led them there.

When the focus stays only on correctness, students may avoid difficult work where the outcome feels uncertain. Naming a strategy changes that message. Instead of saying: "Good job." Try describing the thinking you observed.

For example:

- "I noticed you broke the word into chunks."
- "You reread the paragraph before answering."
- "You used the diagram to double-check your idea."
- "You crossed out weaker evidence and chose stronger support."
- "You paused before responding instead of guessing."

When you name the strategy, the student begins to see what helped them succeed. This shifts confidence from: "I got it right." To "I used a strategy." That difference matters.

When students believe improvement comes from strategy rather than talent, they are more willing to attempt difficult work. They begin to understand that learning is something they can influence through effort, thinking, and revision.

A simple structure can help make strategy naming part of everyday interactions.

Script Formulas

Strategy → Growth

"I noticed you ______. That's a strong strategy because ______."

Examples:

- "I noticed you reread the question before answering. That's a strong strategy because it helps make sure you understand what the problem is asking."
- "I noticed you used the diagram to check your answer. That's a strong strategy because it helps you see if your idea makes sense."

This script shifts the focus from correct answers to effective thinking strategies, helping students recognize the actions that lead to improvement.

Effort → Progress

"You kept ______ even when it was hard. That kind of effort helps your brain learn."

Examples:

- "You kept rereading the problem even when it was confusing. That kind of effort helps your brain figure things out."
- "You kept trying different words until the sentence made sense. That kind of effort helps your reading grow."

This script reinforces persistence and cognitive stamina, which are core confidence builders.

Thinking → Ownership

"Walk me through how you ______ so I can see your thinking."

Examples:

- "Walk me through how you solved that step so I can see your thinking."
- "Walk me through how you decided on that answer."

This script communicates an important message: The student is the thinker.

Instead of evaluating the answer immediately, the adult invites the student to explain their reasoning, reinforcing ownership of the learning process.

Small statements like this help students connect their effort to improvement. You can use this approach regularly. Specific feedback helps students see the relationship between what they do and what they learn.

And over time, that clarity strengthens self-efficacy – the belief that they can grow through their own thinking and effort.

Effort-Specific Praise Guide

Not all praise builds confidence in the same way. Some praise focuses on traits – qualities that sound positive but can unintentionally send the message that success comes from natural ability rather than effort or strategy. For example: "You're so smart." "That was easy for you." "You're a natural."

These comments are well-intended, but they connect success to fixed traits. When students believe their success comes from natural ability, they may become hesitant to attempt work that feels difficult. Struggle can begin to feel like proof that they are no longer "smart" or "talented."

Instead, it is more helpful to reinforce behaviors students can control. These include persistence, revision, question-asking, planning, self-correction, and strategy use. When adults name these behaviors, students begin to understand what actually helps them improve.

Replace Trait Praise With Process Praise

Instead of: "You're good at math." Try: "You kept adjusting your method until it worked."

Instead of: "You're a great reader." Try: "You slowed down when the sentence didn't make sense."

Instead of: "You're talented." Try: "You revised your introduction three times before settling on that version."

These small shifts help students connect success to the actions they took, not to a trait they either possess or lack. Over time, this kind of feedback strengthens a more durable form of confidence.

Confidence built on effort and strategy can survive challenge. Confidence built only on talent often weakens when learning becomes difficult. And learning almost always becomes difficult eventually.

Recovery After Error Protocol

What happens after a student makes a visible mistake often determines whether they will take risks in the future.

Students watch closely to see how adults respond when something goes wrong. If mistakes lead to embarrassment or public correction, students may become cautious. They may stop attempting difficult work or wait for someone else to provide the answer.

But when errors are handled calmly and respectfully, students learn something different: mistakes are part of learning. The goal is not to prevent errors. The goal is to help students recover from them with dignity.

The following steps can help guide that process.

Step 1: Normalize the Struggle

Acknowledge that difficulty is a normal part of learning.

For example: "That's a common place to get stuck."

Statements like this help students understand that confusion does not mean failure.

Step 2: Separate Identity From the Error

Make it clear that the mistake belongs to the work, not to the student.

You might say: "The idea needs revision – not you."

This helps protect the student's identity while keeping the focus on improving the work.

Step 3: Redirect to Strategy

Shift attention toward the next step in the thinking process.

For example: "What could you try next?"

This keeps the student engaged in problem-solving rather than focusing on the mistake itself.

Step 4: Protect Dignity

How help is delivered matters.

Small actions can reduce the feeling of exposure: Lower your voice. Avoid drawing public attention to the error. Stand beside the student rather than above them. These choices help keep the interaction supportive rather than corrective.

Step 5: Close the Loop

Recognize the student's effort to stay engaged.

For example: "I'm glad you stayed with it."

This reinforces persistence and helps the student see recovery as part of the learning process.

Recovery after error often builds more confidence than easy success.

Students rarely remember the exact worksheet or assignment from a lesson. But they do remember how adults responded when they were wrong.

And those moments often shape whether they are willing to try again the next time learning becomes difficult.

Visible Thinking Prompt Cards

One of the most effective ways to support students without taking over their thinking is to rely on simple prompts that invite them to explain, attempt, and reflect.

Some paraeducators find it helpful to keep a small set of thinking prompts in mind throughout the day. These can be written on small cards, kept on a clipboard, or simply remembered as part of your regular support routine.

These prompts encourage students to stay engaged with the work rather than waiting for the answer.

Examples include:

- "What do you think the first step might be?"
- "What strategy could you try here?"
- "What part makes sense so far?"
- "What part feels confusing — specifically?"
- "How would you explain your thinking?"
- "What evidence supports that idea?"

Questions like these slow the interaction just enough to keep the student in the role of the thinker. Instead of immediately explaining the solution, the adult invites the student to process the problem aloud, identify what they understand, and decide what to try next.

This small shift changes the nature of support. The adult moves from rescuer to thinking partner. A helpful guideline is simple: Use questions before explanations. Questions protect ownership.

They help ensure the student remains responsible for the thinking, while still knowing that support is nearby when it is needed.

ELEMENTARY EXAMPLE

Supporting Decoding Without Becoming the Reader

A student struggles with a multisyllabic word. They pause. They look at you.

You could say the word immediately so they can move on. Or you could help them stay in the role of the reader.

Instead, you say:

- "Let's break the word into parts."
- "What sound does this chunk make?"
- "Try blending it slowly."
- "You got the first part – now finish it."

You do not read the word for them. You support the decoding process without replacing it. After a moment, the student blends the sounds and reads the word correctly.

You then name the strategy: "You broke the word into syllables and blended them. That worked."

Now the student knows what helped them succeed.

Confidence increases because effort and strategy produced the result. If you had simply read the word aloud, fluency might improve for that moment. But the student would not experience themselves as the reader.

MIDDLE SCHOOL EXAMPLE

Supporting Lab Analysis Without Giving Interpretations

A student is analyzing lab results. They look over their data and ask: “Is this right?” You could confirm or correct the answer immediately. But doing so would shift the thinking away from the student.

Instead, you guide the analysis. You ask:

- “What pattern do you notice in the data?”
- “What changed between trial one and trial two?”
- “What might explain that difference?”
- “Where do you see evidence for that idea?”

You are not withholding help. You are strengthening the student’s reasoning process. After reconsidering the data, the student revises their explanation.

You respond by naming the thinking: “You connected the temperature change to the reaction rate. That’s scientific reasoning.” The student begins to see what strong analysis looks like.

Confidence grows from reasoning and discovery, not from adult approval alone.

HIGH SCHOOL EXAMPLE

Supporting Essay Planning Without Writing the Thesis

A student sits in front of a blank page. After a long pause, they say: "I don't know what to write." You could suggest a thesis for them so they can begin quickly. But when adults write the thinking, students lose ownership of the argument.

Instead, you guide the thinking process. You ask:

- "What is your position — even loosely?"
- "What are you trying to argue?"
- "What evidence do you already have?"
- "If someone disagreed with you, what might they say?"

Gradually, the student offers an initial idea. You respond: "Let's shape that."

You refine the thinking without replacing it. You might say: "Your claim is about responsibility. Can we make that more specific?"

The student revises the idea and begins writing. You help organize the structure of the essay. But the central claim remains the student's own thinking.

Ownership fuels confidence.

When students know the idea belongs to them, they are far more willing to defend it, revise it, and improve it.

When adults write the thinking for them, the work may look polished – but the confidence underneath it never develops.

Protecting Dignity During Struggle

Struggle is often visible in classrooms.

Students know when they are stuck. They know when the work feels difficult. And when an adult sits beside them, that struggle can feel even more exposed. This is why protecting dignity during these moments matters. Students are far more willing to keep trying when they know their effort will be treated with respect.

You can help protect dignity through small, intentional choices. For example:

- Lower your voice when offering support
- Sit at eye level rather than standing above the student
- Keep facial expressions calm and neutral
- Avoid correcting mistakes publicly
- Avoid sarcasm or humor that might feel personal
- Avoid showing visible frustration when the work is difficult

These small signals communicate safety. When students feel safe, they are more willing to attempt, revise, and try again. Your calm presence also influences how the student experiences the moment.

Emotions are contagious in classrooms. Your calm becomes their regulator.

If you panic, students may internalize that something is wrong. If you rush to fix the problem, they may assume they are incapable. But if you remain steady and patient, students often borrow that steadiness.

And in that calm space, they are far more likely to keep thinking.

IMPLEMENTATION

Confidence Infrastructure Audit

Confidence grows through patterns of experience.

Because paraeducators work closely with students throughout the day, your interactions help shape those patterns. Small moments of support – questions, prompts, encouragement, or pauses – accumulate over time and influence how students see themselves as learners.

To better understand the impact of your support, choose one student you work with regularly and observe your interactions for one week. As you work with that student, track a few simple patterns. Notice:

- How often did I name a strategy the student used?
- How often did I step in and rescue the task?
- How often did I allow productive struggle before helping?
- How often did I protect dignity when the student made an error?
- Did the student attempt the work more independently over time?

At the end of the week, reflect on what you observed. Then ask yourself an important question: Is my proximity building confidence – or comfort?

Comfort removes difficulty quickly. Confidence grows when students experience challenge and discover they can move through it.

Comfort avoids struggle. Confidence survives it.

When support protects dignity while leaving space for thinking, students begin to experience something powerful: the belief that they are capable of learning through effort.

Closing Reflection

Confidence is not something students either have or do not have. It is constructed. Confidence develops through repeated experiences where students attempt something difficult, receive support, revise their thinking, and discover they can improve.

Those experiences accumulate. And you help create many of them. As a paraeducator, you are not "just" support. You are:

- A strategy amplifier
- A protector of dignity
- A regulator of risk
- A partner in thinking
- A builder of learning infrastructure

Each interaction you have with a student helps shape how they experience challenge. You influence whether struggle feels threatening or manageable, whether mistakes feel discouraging or informative, and whether effort feels worthwhile.

In an AI-shaped world, answers are increasingly easy to generate. But courage is not. Students still need the confidence to attempt difficult work, to revise imperfect ideas, and to continue thinking when answers are uncertain.

Your role is to help create the conditions where that courage can grow.

You help students become brave enough to think. And brave enough to try. That kind of confidence is more than a moment of success. It is infrastructure that supports learning long after the classroom lesson has ended.

PART II – BEHAVIOR IS INFORMATION

Chapter 3: Reading What Isn't Being Said

Behavior in classrooms rarely appears without warning. Often there are signals first – small changes that appear before the moment becomes visible to everyone in the room.

Paraeducators are often the first adults to notice those signals. You are closer to the student. You hear the whisper before the outburst. You see the pencil drop before the refusal. You notice the eye roll before the comment. You feel the shift in posture or tone before the teacher does. Your proximity gives you access to information others may not see yet. And what you do in those early moments matters.

Many people think of behavior as something that must be corrected or controlled. But in classrooms, behavior is often communicating something underneath the surface.

Behavior is rarely random. It is information. A student who refuses to begin may be protecting themselves from embarrassment. A student who jokes loudly during difficult work may be covering uncertainty. A student who shuts down may be overwhelmed by the task in front of them.

When behavior is interpreted only as defiance, the response often becomes confrontation. But when behavior is understood as information, the response can shift toward support and regulation. The difference begins with interpretation.

This chapter explores how to recognize what behavior may be communicating, how to respond without escalating the situation, and how to support students in ways that protect dignity while guiding them back into learning. Because what students communicate through behavior is often what they cannot yet say with words.

Behavior Is Communication

Students do not always have the words to explain what they are experiencing internally. When frustration builds, when confusion grows, or when emotions feel overwhelming, behavior often becomes the way those experiences are expressed.

Sometimes the message is loud. A raised voice. A refusal. A disruptive moment. Other times it is subtle. A pencil dropped on the floor. Eyes looking away from the work. A student who suddenly becomes silent. In each case, behavior is communicating something.

When students cannot regulate, explain, or tolerate what they are experiencing internally, behavior speaks.

Researchers in child development and school psychology have long described behavior this way. Dr. Ross Greene, who studies challenging behavior in schools, summarizes it simply: "Kids do well if they can" (Greene, 2014, p.10). When students lack the skills to manage frustration, solve problems, or tolerate difficulty, their behavior often reflects those unmet skills.

Neuroscience research also helps explain this pattern. When students experience stress or threat, the brain prioritizes protection rather than learning. Dr. Bruce Perry explains that in these moments the brain shifts into survival states, which can lead to impulsive reactions, withdrawal, or agitation (Perry & Winfrey, 2021, p. 42). These reactions are not planned misbehavior – they are attempts to regulate overwhelming internal experiences.

This does not mean behavior has no limits or expectations. But it does mean that interpretation matters.

If we interpret behavior only as defiance, the response often becomes confrontation. Voices escalate, corrections become sharper, and the student becomes more defensive. But when behavior is interpreted as information, the response can change.

Instead of reacting immediately, the adult becomes curious. What is the student experiencing right now? Is the work confusing? Is frustration building? Is the student trying to avoid embarrassment?

This shift changes the role of the adult beside the desk.

Your influence is not only in what you say or do. Your influence begins in how you interpret what you see. Your power lies in interpretation.

Not every behavior is willful. Many behaviors are protective. Students may joke to hide uncertainty. They may refuse work to avoid visible mistakes. They may distract others to escape attention. These behaviors do not always mean the student does not care. Often, they mean the student is trying to protect themselves from feeling incapable.

When adults respond with curiosity instead of immediate judgment, the interaction can move in a different direction. Instead of escalating the moment, the adult helps the student regain regulation and return to learning with dignity. And in those moments, behavior becomes not simply a disruption – but information that helps guide the support the student needs.

Four Protective Behaviors You Will See

When students feel overwhelmed, exposed, or uncertain about their ability to succeed, their behavior often shifts. These responses are rarely random. They are attempts to protect identity, reduce embarrassment, or regain a sense of control.

Researchers who study student behavior often describe these reactions as protective responses – ways students attempt to manage stress or perceived threat (Perry & Winfrey, 2021, p. 42; Greene, 2014, p. 10). Recognizing these patterns can help adults respond with support rather than escalation.

Below are four protective behaviors you may notice in classrooms.

1. Humor as Armor

Sometimes the first response to difficulty is humor. You might notice: Jokes when work becomes difficult. Sarcasm when corrected. Acting like the class clown during serious moments.

Humor can function as protection.

If the student can make everyone laugh, attention shifts away from the work they may not understand. Humor can quietly communicate: "Don't look too closely." The student protects their identity by turning the moment into entertainment rather than exposure.

2. Silence as Protection

Other students respond to difficulty by becoming very quiet. You might notice: No response when called on. Shrugging shoulders. Looking down at the desk. Minimal engagement with the task.

Silence can protect vulnerability.

If the student does not respond, they cannot be visibly wrong. Silence can communicate: "I'm not safe enough to risk." Students often withdraw verbally when they believe their answer may lead to embarrassment.

3. Defiance as Overwhelm

Some behaviors appear confrontational. You might see: Refusal to begin the task. Pushback toward directions. An argumentative tone. Statements such as "I'm not doing this.".

While defiance can sometimes be intentional, it is often connected to overwhelm. When students feel incapable of completing a task, resisting the task can restore a sense of control (Greene, 2014, p. 17). In these moments, the behavior may be communicating: "This feels too big." The student is attempting to escape a situation where they expect failure.

4. Withdrawal as Exposure Risk

Other students respond by quietly removing themselves from the situation. You might notice: Head down on the desk. Very slow work or delayed start. Frequent bathroom or nurse requests. Attempts to leave the room.

Withdrawal reduces exposure.

If the student leaves the situation, their confusion is no longer visible to others. Withdrawal can communicate: "I don't want my struggle to be seen." Students often disengage when they believe their difficulty may be witnessed by peers.

Understanding these behaviors does not mean ignoring expectations. But it does mean recognizing that behavior often reflects something happening beneath the surface.

When adults interpret behavior as information, they can respond in ways that protect dignity while guiding the student back toward learning.

The Key Distinction

Paraeducators often encounter these behaviors before they become visible disruptions in the classroom. You are closer to the student. You notice the signals earlier. You see the moment when frustration begins to rise.

Because of this proximity, you often become the first interpreter of what is happening. And interpretation matters. How you understand the behavior influences what happens next.

Your interpretation can shape whether the moment becomes: Escalation or de-escalation. Shame or safety. Correction or curiosity. Dependence or growth.

When behavior is interpreted only as disrespect, the response often becomes sharp and corrective. But when behavior is understood as protection, the response can become more thoughtful and strategic.

A joking student may be protecting uncertainty. A silent student may be protecting vulnerability. A resistant student may be protecting themselves from feeling incapable.

Recognizing this does not mean ignoring expectations. It means responding in a way that keeps the student's dignity intact while guiding them back toward learning.

Behavior is information.

Your role is to notice it, interpret it carefully, and respond in ways that help students recover and continue thinking.

When adults read behavior wisely, many disruptions can be softened before they ever fully emerge.

Behavior Reframing Guide

When challenging behavior appears, the first reaction is often quick judgment. But a brief pause can help shift the response from reaction to understanding.

Instead of responding immediately, take a moment to ask a few guiding questions. Consider:

1. What might this student be protecting right now?

2. What is the cognitive demand of this task?

3. Has the difficulty level increased recently?

4. Is the student being publicly exposed in this moment?

5. Is this behavior new, or part of a pattern I have seen before?

These questions help move the focus from blame to interpretation.

Reframing behavior does not excuse the behavior itself. Expectations still matter. Students still need guidance and boundaries. But reframing helps ensure the adult response addresses the underlying need, not just the visible action. For example:

Replace the thought: "He's being disrespectful." With: "He may be protecting himself."

Replace: "She doesn't care." With: "She may not feel capable right now."

Replace: "They're lazy." With: "They may be overwhelmed."

These shifts in thinking help adults respond with calm, curiosity, and strategy rather than immediate frustration.

Reframing does not remove accountability.

It refines the response so that correction and support can happen without damaging the student's dignity.

Armor vs. Disrespect Checklist

When challenging behavior appears, it can be tempting to react immediately. But pausing briefly to observe patterns can help determine what the behavior might be communicating.

Before escalating the situation, consider the context in which the behavior is occurring. Ask yourself:

☐ Is the behavior happening mainly during challenging tasks?
☐ Does the behavior increase when attention becomes public?
☐ Does it decrease when the work becomes easier or more familiar?
☐ Has the student shown effort in lower-risk situations?
☐ Does humor appear after moments of confusion?
☐ Does silence appear after correction or feedback?
☐ Does refusal occur when multi-step directions are given?

If most of these answers are yes, the behavior may be functioning as armor rather than intentional disrespect. In these moments, the student may be protecting themselves from embarrassment, confusion, or the fear of appearing incapable.

However, if the behavior appears consistently across many situations – including during preferred activities or low-demand tasks – it may signal a different pattern that requires teacher or administrative support.

Paraeducators are not responsible for diagnosing intent or determining discipline responses. But you are well positioned to notice patterns.

Your proximity allows you to observe when behaviors appear, what triggers them, and how students respond to different types of support.

Pattern recognition helps adults respond more thoughtfully.

And thoughtful responses help protect student dignity while guiding them back toward learning.

Exposure Risk Guide

Some classroom tasks carry a higher level of exposure than others.

Exposure refers to how visible a student's thinking becomes in front of others. When students believe their uncertainty or mistakes may be seen publicly, the emotional risk of participating increases.

As exposure rises, the likelihood of protective behaviors – such as humor, avoidance, or refusal – often rises as well.

To anticipate these moments, consider how exposed the student may feel during certain activities. Use a simple scale: 1 = Low exposure to 5 = High exposure.

Rate common classroom tasks such as:

- Reading aloud in front of the class
- Explaining reasoning publicly
- Writing answers on the board
- Sharing an unfinished or uncertain response
- Participating without preparation
- Being corrected publicly

When a task feels like a 4 or 5 on the exposure scale, students may experience higher stress or hesitation.

In these moments, it can be helpful to reduce exposure before difficulty turns into disruption.

Instead of waiting for behavior to escalate, consider small adjustments that make the task feel safer. You might:

- Offer quiet rehearsal before participation
- Preview vocabulary or directions privately
- Allow a written response before asking for an oral one
- Sit beside the student rather than standing over them
- Provide additional think time before responding

These adjustments do not lower expectations. They lower the public risk while the student is still learning.

When exposure is reduced, students are more likely to attempt the work rather than protect themselves from it.

Lower exposure before correction. Prevention helps protect confidence.

Correction vs. Curiosity Script

When challenging behavior appears, the tone of the adult response can influence what happens next. If the tone communicates threat or accusation, the situation often escalates. Students may become defensive, shut down, or push back more strongly. But when the tone communicates curiosity and partnership, the interaction often slows down. Students feel less exposed and more willing to explain what is happening.

Correction focuses on stopping the behavior immediately. Curiosity focuses on understanding what the behavior may be communicating. For example, instead of saying: "Stop that." "That's disrespectful." "You need to behave." "You know better than that." Try beginning with observation and curiosity.

You might say: "I'm noticing you're joking right now. What's going on?" "This seems frustrating. What part feels stuck?" "I can see this is hard. Tell me what's confusing." "Are you feeling overwhelmed?" Statements like these communicate that the adult is trying to understand the situation rather than immediately punish the behavior.

This does not mean expectations disappear.

Students still need guidance and boundaries. But curiosity helps ensure the response addresses the reason behind the behavior, not just the behavior itself.

Curiosity does not mean permissiveness. It means investigation before instruction.

Once the purpose of the behavior becomes clearer, the adult can respond more effectively – supporting the student while still guiding them back toward appropriate behavior and learning.

Elementary Example

Reading What Isn't Being Said During Independent Reading

During silent reading, a third grader suddenly begins giggling loudly. Other students notice. A few begin laughing as well.

From the outside, it may look like the student is simply being disruptive.

You could respond quickly with: "Stop being disruptive." But instead, you move closer and kneel beside the student. In a quiet voice, you ask: "Is the text feeling tricky right now?" The student shrugs.

Rather than calling attention to the moment, you point gently to the page. "Let's read this sentence together quietly."

As you read together, you notice the student struggling with several multisyllabic words. The giggling had begun at the moment the decoding broke down.

Humor was functioning as armor.

By noticing the skill difficulty rather than judging the behavior, you protect the student's dignity and help them re-engage with the task.

You addressed the reading challenge, not the student's character.

Middle School Example

Lab Group Refusal

During a science lab, a seventh grader refuses to record the group's data. They push the clipboard away and say: "I'm not writing that."

At first glance, the behavior may appear defiant.

You could respond with correction: "You need to participate with your group." Instead, you pause and ask a question that focuses on the task. "This is the analysis section. Is the interpretation part feeling unclear?"

The student hesitates, then nods slightly.

You continue: "What pattern are you noticing in the results?"

As the conversation shifts toward the data itself, the student begins to re-engage. They look back at the experiment results and slowly start contributing ideas.

The refusal was not simply attitude. It was overwhelm with the cognitive demand of the task.

By addressing the difficulty of the analysis rather than confronting the behavior alone, you helped the student regain confidence and participate again.

High School Example

Essay Discussion Silence

During a class discussion about essays, students are asked to share their ideas with peers.

One student sits quietly. Head down. Minimal eye contact. No participation.

From a distance, it might appear as disengagement. But when the group activity pauses, you quietly check in. “Do you want to rehearse your idea before sharing it with the group?”

The student looks up and quietly explains a thoughtful interpretation of the text. You respond: “That’s a strong point. Want me to help you phrase it so it’s easier to share?” After practicing briefly, the student joins the discussion.

In this moment, silence was not indifference. It was exposure risk.

The student had an idea but did not feel ready to share it publicly.

By lowering the exposure threshold, you helped the student move from hesitation to participation.

And in that moment, confidence had room to grow.

Teaching as Design Reminder

Behavior in classrooms is not separate from instruction. The way lessons are structured, the expectations around participation, and the emotional tone of the classroom all influence how students respond.

Behavior is often shaped by the conditions surrounding the learning task. These conditions include: task difficulty, public exposure during participation, the timing and pacing of activities, patterns of feedback and correction, the emotional climate of the classroom, and the predictability of routines and expectations.

When students begin to show signs of frustration, avoidance, or resistance, it can be helpful to look beyond the behavior itself. This is a time to ask a broader question: What in the learning environment might be contributing to this moment?

Sometimes the task may be too complex without enough support. Sometimes the activity may require public participation before students feel prepared. Sometimes uncertainty about expectations can increase anxiety.

These conditions influence how students experience the work.

Paraeducators are not responsible for designing the lesson or determining the instructional plan. But you do play an important role in how behavior is interpreted and responded to in the moment.

Your proximity allows you to notice early signals that others may not see yet. Because you are close to the student, you often recognize frustration, confusion, or hesitation before it becomes a visible disruption. In that sense, you become an early warning system for behavior.

When these signals are noticed early, small adjustments – quiet clarification, reassurance, or a brief check-in – can often prevent larger disruptions later. Your proximity is powerful. Use it thoughtfully to interpret behavior, protect dignity, and help students return to learning.

IMPLEMENTATION

Behavior Response Audit

The way adults interpret behavior often shapes what happens next.

Because paraeducators are frequently the first adults to notice early signals – frustration, avoidance, humor, or refusal – your response can either calm the moment or unintentionally intensify it.

To better understand the impact of your responses, try a brief observation exercise. Choose one week and reflect on your interactions with students when challenging behavior appears.

Track the following patterns:

- How many times did I correct the behavior immediately?
- How many times did I begin with a curiosity question instead?
- Did behaviors escalate less often when I used curiosity?
- Did the student's tone or posture shift after I reframed the behavior?
- Was the student's dignity preserved during the interaction?

At the end of the week, pause and reflect. Ask yourself: Did my interpretation help calm the situation, or did it intensify it?

Often the difference between escalation and recovery begins with how the adult understands the behavior. A quick assumption of disrespect can trigger confrontation. A moment of curiosity can create space for understanding. Remember: Escalation often begins with adult interpretation.

Dignity often begins with adult restraint. When adults slow down long enough to interpret behavior thoughtfully, many difficult moments can return to learning before they grow into larger disruptions.

Closing Reflection

Paraeducators often notice behavior first. This is not accidental.

Because you work closely with students, you are often positioned near the moments when frustration, uncertainty, or overwhelm begins to appear. You see the early signals – before the behavior becomes visible to the entire classroom.

In many ways, you are positioned closest to student vulnerability. What happens next often depends on how that moment is interpreted.

Interpretation can lead in different directions. A moment of vulnerability can turn into shame, escalation, or removal from the learning environment. Or it can lead to support, regulation, and growth. The difference often begins with how the adult understands what they are seeing.

Behavior is information. It tells us when a task feels overwhelming, when exposure feels risky, or when a student is trying to protect themselves from embarrassment.

When behavior is interpreted carefully, adults can respond in ways that protect dignity while guiding students back toward learning. And when dignity is protected, students are far more willing to remain engaged.

Because learning requires risk. And students are most willing to take those risks in environments where their dignity remains intact.

Chapter 4: Supporting Without Creating Dependence

This may be one of the most important ideas in this book.

Support can strengthen independence. Or it can quietly weaken it. The difference is often difficult to notice in the moment.

Many interactions between students and support professionals are brief – just a few seconds beside a desk, a quick answer to a question, or a short explanation during independent work.

But those small moments accumulate. Over time, they shape how students respond when learning becomes difficult.

Consider a familiar situation. A student looks up from their work and asks: "Can you help me?" At first, the question seems simple. But what happens next matters. If help arrives immediately, the student may begin to rely on the adult whenever the task becomes challenging. If the response encourages thinking first, the student may begin to see themselves as capable of working through difficulty.

In other words, the way support is offered influences how students experience challenge. Do they look outward for rescue? Or do they begin to look inward for strategies?

Paraeducators encounter this moment many times throughout the day. Each time a student asks for help, the adult beside the desk has an opportunity to guide the student toward greater independence.

This chapter explores how to provide meaningful support while still protecting the student's role as the thinker. Because the goal of support is not simply helping students complete the work in front of them. The goal is helping them develop the confidence to continue thinking when the work becomes difficult.

The Risk of Learned Helplessness

Support is essential in classrooms.

Many students need guidance, clarification, and encouragement as they learn new skills. Paraeducators play an important role in providing that support. But there is an important balance. When help arrives too quickly or too frequently, students may begin to rely on the adult rather than on their own thinking.

Researchers describe this pattern as learned helplessness. Psychologist Martin Seligman first used this term to describe situations where individuals stop attempting solutions because they believe their actions will not change the outcome (Seligman, 1975). In school settings, this can appear when students wait for help before beginning work or avoid attempting tasks independently.

Over time, learned helplessness can develop through repeated experiences such as:

- Adults stepping in before the student attempts the task
- Someone else resolving confusion immediately
- Struggle being interrupted quickly
- Adults showing discomfort when the student experiences difficulty
- Rare expectations for independent problem-solving

When these patterns occur repeatedly, students begin to internalize a different message about themselves as learners.

The internal dialogue may sound like: "I can't do this." "I shouldn't try." "I need someone to help me." "I won't start unless someone is beside me."

Carol Dweck's research on motivation shows that when students interpret difficulty as evidence that they lack ability, they are more likely to withdraw effort and stop trying (Dweck & Diener, 1980). Without opportunities to attempt, revise, and recover from mistakes, students may never develop the belief that effort can lead to improvement.

This does not mean paraeducators cause learned helplessness.

Support is necessary, and many students benefit greatly from adult guidance. But proximity without intentional fading can unintentionally create dependence.

When adults remain beside a student constantly, the student may come to believe that learning only happens when the adult is present.

Albert Bandura's research on self-efficacy explains why this matters. Students develop confidence when they experience themselves solving problems and overcoming difficulty through their own effort (Bandura, 1997). When those experiences are removed, confidence has fewer opportunities to grow.

The goal of support is not simply helping students finish the task. The goal is helping them become stronger thinkers.

Support should gradually build independence over time. It should strengthen the student's belief that they can attempt, struggle, revise, and succeed. Because when students experience themselves solving problems, something important changes: They begin to believe they are capable.

Over-Scaffolding

Scaffolds are an important part of learning. They help students begin tasks that might otherwise feel too difficult. Sentence starters, step-by-step guides, visual supports, and adult prompts can all make challenging work more manageable. But scaffolds are meant to be temporary supports, not permanent structures.

Their purpose is to help students get started until they can take more ownership of the task. In classrooms, however, scaffolds sometimes remain in place longer than intended. Sentence starters may appear on every assignment. Step-by-step guides may never be removed. An adult may remain beside the same student during every independent task.

When scaffolds stay in place indefinitely, they can unintentionally change the learning process. Over time, students may begin to rely on the scaffold rather than on their own thinking. When scaffolds do not fade, thinking becomes weaker, risk tolerance decreases, initiative begins to drop, and dependence on adults increases.

Students may still complete the work. The assignment may be finished, the worksheet filled in, or the answer recorded. But completion is not the same as growth.

For learning to deepen, students need opportunities to attempt tasks with gradually less support. Scaffolds work best when they are designed to fade.

As students gain confidence and skill, supports should slowly step back so that the student's own thinking can take a larger role.

The goal of scaffolding is not to hold students up forever. The goal is to help them stand on their own.

The Support Spectrum

To help students grow more independent, it is useful to think about support along a continuum. Not every moment requires the same level of assistance. Sometimes students need direct help to begin. At other times, the most supportive action is stepping back so they can attempt the work on their own.

One way to think about this is through The Support Spectrum. Every interaction between an adult and a student falls somewhere along this range of support.

1. Do for the student
2. Do with the student
3. Prompt the student
4. Question the student
5. Observe silently

The key skill for paraeducators is recognizing when to shift levels. Support that begins strong should gradually move toward independence whenever possible.

Level 1: Do for the Student

At this level, the adult completes the task. You may write the sentence. Solve the equation. Read the passage aloud. Fill in the response. This level should be used sparingly.

It is most appropriate in situations such as: Significant cognitive barriers. Crisis or emotional regulation moments. IEP-directed accommodations. Situations involving immediate safety. While this level may help the student complete the task, it does not build independence. It provides comfort, but it does not strengthen the student's role as the thinker.

Level 2: Do with the Student

At this level, the adult works alongside the student. You model the process and complete part of the task together. You might say: "Let's write the first sentence together." "Let's solve this first step together."

This level is especially helpful when a student is learning a new skill. The adult provides structure while the student participates in the process. However, the thinking is still heavily supported by the adult. Independence at this level is shared.

Level 3: Prompt the Student

At this level, the adult begins stepping back. Instead of completing the task together, you offer cues that help the student recall strategies or next steps. You might ask: "What comes next?" "Look at your notes." "What strategy did we use yesterday?"

Prompts help reduce cognitive overload while still keeping the student responsible for the thinking. At this stage, independence begins to strengthen.

Level 4: Question the Student

Here, the student takes primary ownership of the work. The adult's role shifts to asking questions that encourage thinking and explanation. For example: "What do you think?" "Why does that work?" "How would you start?" "What evidence supports that idea?"

At this level, the adult becomes a thinking partner rather than a rescuer. The student remains responsible for solving the problem.

Level 5: Observe Silently

This level is often the most difficult.

Instead of stepping in immediately, the adult watches and allows the student to continue working independently. You observe. You wait. You allow productive struggle to unfold.

For many adults, this level requires patience and trust.

Silence can feel uncomfortable, especially when a student hesitates or pauses. But silence is not neglect. It is confidence in the student's ability to attempt, think, and grow.

When students are given space to work through challenges, they begin to experience something powerful: They discover they can do more than they thought.

The Critical Skill: Shifting Levels

Providing support is not only about choosing the right level of help. It is also about knowing when to shift that level over time.

In classrooms, adults often remain at the same level of support longer than intended. This is especially common at the first two levels of the spectrum.

Paraeducators may find themselves repeatedly completing tasks for the student or working through every step together. These forms of support can be helpful in the moment. They allow the student to begin the task and reduce frustration. But when support stays at these levels for too long, the student may begin to rely on the adult rather than developing their own strategies.

Independence grows when support gradually moves upward along the spectrum. The goal is steady movement in this direction:

Do for → Do with → Prompt → Question → Observe

For example, you might begin by working through the first part of a task together. Once the student understands the structure, you move to prompts that remind them of the next step. As confidence increases, prompts can shift into questions that encourage the student to explain their thinking. Eventually, the adult steps back and observes while the student works independently. Movement along this spectrum helps students experience increasing ownership of the work.

If support begins at Level 2, it should not remain there indefinitely. Growth requires movement. Each shift in support helps students take on a larger role in the thinking process. And over time, those shifts build the independence that allows students to approach challenges with greater confidence.

Prompt Fading Tracker

Independence develops gradually.

One way to support that growth is to notice how often different levels of support appear in your interactions with students. This short reflection tool can help you observe your patterns.

Choose one student you support regularly and track your interactions for one week. During that week, note how often each level of support occurs.

- How often did I do the task for the student?
- How often did I do the task with the student?
- How often did I prompt the student with cues or reminders?
- How often did I ask questions that encouraged thinking?
- How often did I observe quietly while the student worked independently?

At the end of the week, take a moment to reflect on what you noticed. Ask yourself: Is my support gradually moving upward along the support spectrum, or remaining at the same level? Ideally, support shifts over time toward greater independence.

If most interactions remain at Levels 1 or 2, the student may be receiving consistent help but not developing greater ownership of the work.

This reflection is not meant to create pressure or self-criticism. Instead, it helps highlight opportunities to gradually step back so the student can take a larger role in the learning process.

Small shifts in support can lead to meaningful growth in independence.

"Three Before Me" Protocol

Students often ask for help quickly when a task becomes confusing. In many classrooms, this request happens before the student has had an opportunity to attempt the work independently.

One way to build independence is to create a simple routine that encourages students to try a few strategies before asking for adult support. For example, students might be encouraged to try three steps first:

1. Reread the directions carefully.

2. Use a posted strategy, example, or class notes.

3. Ask a peer or make an independent attempt.

Only after these steps do they ask the adult for help.

This structure helps students pause and attempt problem-solving before relying on outside support. You might gently remind a student: "I want to see three attempts before I step in." This statement does not deny help. Instead, it communicates an expectation that the student's thinking comes first.

Over time, many students begin to try these strategies automatically. They reread the directions. They look for examples. They attempt a first step before seeking assistance. When this happens, the student is beginning to take ownership of the learning process. And that shift – trying before asking – is an important sign of growth.

Independence Growth Map

Independence rarely develops all at once. More often, it grows gradually as students gain confidence and experience solving problems with decreasing levels of support.

One helpful way to notice this growth is to track how a student's need for support changes over time. Choose a student you work with regularly and observe how their level of independence develops across several weeks.

For example:

Week 1: The student needs step-by-step modeling to begin the task.

Week 3: The student can begin with prompting or reminders.

Week 6: The student starts independently but benefits from occasional check-ins.

Week 10: The student begins tasks without prompting and works with greater confidence.

These time frames will vary depending on the student, the skill being learned, and the complexity of the task. The goal is not speed, but movement toward greater independence. Over time, the level of support should gradually decrease as the student's confidence and competence grow.

If weeks pass and you do not notice meaningful changes in independence, it may be helpful to pause and reflect. Ask yourself: Is my support gradually fading, or am I providing the same level of help each time?

Independence should grow across weeks and months. If it does not, the support structure may need to be redesigned so that the student has more opportunities to attempt, struggle productively, and take ownership of the work.

The Independent Release Model

Many teachers organize instruction around a simple framework often described as: "I do, we do, you do." First, the teacher models the skill. Then the class practices together. Finally, students attempt the task independently.

Paraeducators play an important role in supporting this same progression. For paraeducators, the process can be understood as: I do for → We do together → You do with prompts → You do independently. Each step gradually shifts responsibility from the adult to the student.

In the beginning, the student may need direct modeling. The adult shows how to approach the task. Next, the student works alongside the adult, completing parts of the task together. As confidence grows, the adult begins offering prompts or questions rather than direct assistance. Eventually, the student works independently while the adult observes and checks in when needed.

One common challenge in this process is remaining at the "we do" stage for too long. Working together often feels comfortable. It is efficient. It keeps the task moving. It reduces frustration in the moment. But when support stays at this level indefinitely, the student may not have the opportunity to develop independence.

Gradual release requires movement. And sometimes that movement feels uncomfortable – for both the adult and the student. There may be moments when the student hesitates or struggles. In those moments, the natural instinct is to step in quickly. Instead, pause. Ask a question. Allow the student a moment to attempt the next step. Then step back and observe.

These small pauses create space for the student's thinking to grow. And over time, those moments help students discover that they are capable of doing more on their own.

Elementary Example

Math Problem Solving

A fourth-grade student looks at the worksheet and waits for you to solve the first problem. The pencil stays still.

They glance at you, expecting help. You could complete the first step for them so the work begins quickly. Instead, you ask: "What do you think the first step might be?"

The student shrugs. Rather than solving the problem, you redirect their attention. "Look at yesterday's example."

They scan their notes and begin attempting the first step. You resist the instinct to reach for the pencil.

After a moment, the student writes the first step on their own.

Now you shift your support slightly. "Why did that step work?" The student explains their reasoning.

Then you step back again. You observe as the student begins the second problem independently.

In this moment, growth occurred because the support gradually faded.

The student remained the problem solver.

Middle School Example

Science Analysis

During a science activity, a student studies a graph and says: "Just tell me what this graph means." You could quickly explain the pattern so the assignment continues.

Instead, you respond with a question. "What pattern do you notice in the graph?"

The student looks at the page. Silence. You wait.

After several seconds, the student says quietly: "The line goes up."

You follow with another question. "What might that increase indicate?"

The student begins explaining possible interpretations of the data.

Gradually, the explanation becomes clearer and more detailed.

If you had explained the graph immediately, the assignment might have been completed more quickly. But the thinking would not belong to the student.

By asking questions and allowing wait time, you helped the student construct the meaning themselves.

High School Example

Essay Organization

A high school student hands you a blank outline. "I don't know how to structure this essay." You could quickly rearrange their ideas or suggest a thesis.

Instead, you begin with a question. "What is your main argument?"

The student answers loosely. You continue: "What evidence supports that idea?"

They list two pieces of evidence. You ask: "Where might those fit in your outline?"

The student begins organizing the structure.

You resist touching the keyboard or writing on the paper.

The ideas remain the student's. The organization begins to take shape.

In this moment, the student is not simply completing an assignment.

They are practicing how to structure their own thinking. And because the work remains theirs, ownership – and confidence – remain intact.

The Discomfort of Stepping Back

It is important to say this clearly: Stepping back can feel risky.

When a student hesitates or struggles, the natural instinct is often to step in quickly and make the moment easier. Paraeducators care deeply about students, and it can be uncomfortable to watch a student wrestle with a difficult task.

You may find yourself thinking: "They might fail." "They might shut down." "The teacher may think I'm not helping." "They look uncomfortable." These reactions are understandable. Adults often want to reduce discomfort for students as quickly as possible. But learning rarely happens without some level of challenge.

Short-term discomfort can help build long-term strength.

When students experience manageable struggle and work through it successfully, they begin to trust their ability to handle difficulty. If discomfort is always removed immediately, students may never develop that confidence.

Independence grows when students are given opportunities to attempt, think, and recover from mistakes. This does not mean leaving students unsupported. It means remaining nearby while allowing space for the student's thinking to develop.

Your calm presence communicates something important. It signals: "I believe you can handle this." And that message often gives students the confidence to keep trying.

When to Step Back Down the Spectrum

The Support Spectrum is meant to guide decisions, not create rigid rules.

Students will not move smoothly from one level to the next every time. Some moments of learning are steady. Others include hesitation, confusion, or brief overwhelm.

When that happens, it may be appropriate to step down one level on the spectrum temporarily.

For example, if a student becomes frustrated while working independently, you might shift from observing silently to asking a guiding question.
If questions are not enough, you might offer a prompt or reminder of a strategy.

These adjustments help the student regain footing without removing ownership of the task.

The important principle is this: Support may move down the spectrum briefly, but it should not stay there.

Once the student regains momentum, begin shifting upward again:

Do for → Do with → Prompt → Question → Observe

The overall direction of support should always trend toward greater independence.

Temporary assistance can help students recover from difficulty.

Permanent assistance can unintentionally create dependence.

The goal is not to remove support. The goal is to ensure that support continues to fade as students grow stronger in their own thinking.

IMPLEMENTATION

One Student, One Shift

Building independence does not require changing everything at once. Often, meaningful growth begins with one intentional adjustment in how support is provided.

Choose one student you work with regularly and reflect on your current pattern of support. Ask yourself: Which level of the Support Spectrum do I most often operate at with this student? Do I usually do the task for them, work alongside them, offer prompts, ask questions, or mostly observe while they work?

Once you identify the level you most commonly use, try a small shift. For the next two weeks, intentionally move one level higher on the Support Spectrum whenever possible. For example if you often do the work with the student, shift toward prompting them instead. If you usually prompt, shift toward asking questions that encourage thinking. If you frequently ask questions, try allowing more moments of silent observation.

As you try this shift, notice what changes. Track questions such as:

- Did the student attempt the work more often?
- Did the number of help-seeking questions decrease?
- Did the student begin initiating steps independently?
- Did frustration increase briefly at first?
- Did the student recover more quickly after moments of confusion?

These reactions are normal. When students are used to a certain level of support, even a small shift can feel unfamiliar at first. Growth may look messy before it looks strong. But over time, these small shifts help students develop the confidence to attempt, persist, and think more independently.

Final Reflection

Supporting students without creating dependence is a thoughtful practice. It requires attention and balance.

Effective support often involves:

- Restraint when the instinct is to step in quickly
- Intentional fading of assistance over time
- Strategic questions that keep the student thinking
- Emotional steadiness during moments of struggle

Paraeducators do more than help students complete assignments. You help students develop the confidence and skills to continue learning when help is not immediately available.

In many ways, your work prepares students to function without constant adult guidance. That is not a loss of support. It is a sign of growth.

The goal of support is not permanent closeness. The goal is capacity.

Capacity means students gradually learn to attempt, think, revise, and recover on their own.

Over time, the strategies and confidence you help build begin to travel with them. One day, they may not need you beside the desk. But the habits of thinking, persistence, and independence you helped develop will remain. And that is the lasting success of support.

PART III – THINKING MATTERS MORE THAN ANSWERS

Chapter 5: Questions You Can Use

Paraeducators often ask a simple question: "What am I supposed to say?" It is an understandable concern.

Paraeducators are not typically responsible for designing the lesson. You are not selecting the standards, planning the assessment, or determining the instructional sequence for the class. But you are often present at one of the most important moments in learning. You are standing beside the student while they are thinking.

In those moments, what you say can shape what happens next. A quick explanation may end the struggle. A well-timed question may extend the thinking. Both move the moment forward, but they lead to very different outcomes.

When answers arrive too quickly, the task may be completed, but the thinking disappears. When questions invite the student to explain, attempt, or reconsider an idea, the student remains the thinker.

Paraeducators may not design the lesson. But you often help shape how students experience it. The words you choose in those brief interactions beside the desk can influence whether students rely on answers or develop their own thinking. Support is not about having the correct answer ready. It is about protecting the student's opportunity to think.

This chapter explores simple questions you can use in everyday classroom moments – questions that keep the student engaged in the work while still providing support. Because sometimes the most powerful support is not an explanation. It is the right question at the right time.

The Shift: From Answer Provider to Thinking Partner

Students often ask a familiar question: "What's the answer?" On the surface, the request seems straightforward. But in many cases, the student is not only asking for the answer. They are asking for relief from the discomfort of uncertainty. They may be thinking: "This feels confusing." "I don't want to be wrong." "I don't know where to start."

When an adult provides the answer immediately, the discomfort disappears. But something else disappears as well. When answers arrive too quickly, the student loses the opportunity to experience ownership of the task, strategy use, productive struggle, memory formation, and confidence building.

The work may be completed. But the learning process becomes weaker. The goal of support is not to deny help. Students still need guidance, encouragement, and clarification. The goal is to redirect the moment toward thinking.

Questions help accomplish this shift. A well-timed question can move the student back into the role of the thinker. Effective questions often help to:

- Return ownership of the task to the student
- Activate strategies the student already knows
- Surface the student's reasoning
- Slow down guessing and rushing
- Protect the student's dignity while they attempt the work

When questions guide the interaction, the student remains responsible for the thinking. The adult becomes a thinking partner rather than an answer provider. You are not there to supply every answer. You are there to protect the process that allows students to develop their own.

Core Replacement Pattern

When students ask for the answer, the goal is not to refuse help. The goal is to guide the student back into the thinking process. One helpful approach is to replace answer-giving with questions that prompt the student to reflect on what they already know or what they have already attempted.

Instead of responding to: "What's the answer?" Try asking questions that redirect the student's attention toward the task. For example:

- "What strategy did your teacher show you?"
- "What part makes sense so far?"
- "What do you notice?"
- "Where did you get stuck?"
- "What might the first step be?"
- "What have you tried already?"

These questions help slow the interaction and shift the focus back to the student's thinking. They also send an important message: the student is capable of engaging with the problem. When used consistently, questions like these accomplish three important things.

First, they reduce panic. Students often feel overwhelmed when they believe they must produce the correct answer immediately. Questions help break the task into manageable steps. Second, they signal belief. By asking questions rather than supplying answers, you communicate that you trust the student's ability to think through the problem. Third, they restore ownership. The responsibility for solving the task returns to the student rather than shifting to the adult. In this way, questions become a simple but powerful tool.

Questions help keep the student in the role of the thinker while still providing the support needed to move forward.

Thinking-Protecting Prompts

Questions can help keep students engaged in the thinking process, especially when a task becomes difficult. Certain prompts work across grade levels and subjects because they focus on helping students clarify what they know, identify their next step, and explain their reasoning.

These questions do not replace instruction. Instead, they help students stay connected to the work while support is nearby. Below are examples of prompts that protect the student's role as the thinker.

When a student feels stuck

Sometimes students pause because they are unsure where to begin. You might ask:

- "What do you know for sure so far?"
- "What might the first small step be?"
- "Where does it start to feel confusing?"
- "Have you seen a similar problem before?"

These prompts help students break the task into smaller pieces and reconnect with what they already understand.

When a student is guessing

Students sometimes offer quick answers without fully thinking through the problem. Questions can slow the process and encourage reasoning. You might ask:

- "How do you know?"
- "What evidence supports that idea?"
- "Can you explain your reasoning?"

These prompts encourage students to examine their thinking rather than relying on guesswork.

When a student begins to shut down

When a task feels overwhelming, students may stop engaging altogether. In these moments, it can help to reduce the demand slightly while still keeping the student involved. You might say:

- "Let's just try the first piece."
- "What part feels manageable right now?"
- "Would you like to think out loud together?"

These questions acknowledge the difficulty while helping the student take a small step forward.

When a student asks for the answer

Requests for answers often come when a student feels uncertain or frustrated. Rather than supplying the answer immediately, you can redirect the student back into the process. For example:

- "Show me what you've tried so far."
- "Talk me through your thinking."
- "What strategy might help here?"

These prompts encourage the student to remain responsible for the work while still receiving support.

Thinking-protecting prompts help ensure that the task stays in the student's hands. The adult remains nearby as a guide, but the thinking remains with the learner.

"Explain Your Thinking" Scaffolds

Students are often asked to explain their thinking. But for many learners, the challenge is not only solving the problem. It is finding the words to describe how they solved it. Without support, the request to "explain your thinking" can feel overwhelming. Students may respond with a short answer, a shrug, or silence. Simple scaffolds can make this process easier.

Instead of repeating the same instruction, provide language structures that help students organize their explanation.

Sentence Frames

Sentence frames give students a starting point for describing their reasoning.

For example:

- "First, I ______."
- "I noticed that ______."
- "I chose this because ______."
- "The evidence shows ______."
- "I got stuck when ______."
- "I changed my answer because ______."

These frames help students focus on the steps they took and the reasoning behind those steps.

Structured Prompts

Another approach is to break the explanation into smaller pieces.

You might guide the student by saying:

- "Tell me what you did."
- "Tell me why you did it."
- "Tell me how you know it works."

These prompts help students slow down and reflect on the process they used.

Over time, students become more comfortable explaining their reasoning. And explanation plays an important role in learning.

When students explain their thinking, they organize their ideas, notice gaps in their reasoning, and deepen their understanding.

When students explain the idea themselves, they begin to own it.

When adults explain everything for them, students may understand temporarily – but the thinking does not fully belong to them.

AI-Resistant Support Questions

In today's classrooms, answers are easier to obtain than ever before. Students can often generate explanations, summaries, or solutions quickly with digital tools. But the presence of answers does not guarantee understanding. What still requires effort is thinking – interpreting information, weighing evidence, and explaining why an idea makes sense.

Paraeducators can help protect these deeper thinking processes by asking questions that require students to reason rather than simply retrieve information. Examples include:

- "Why do you believe that?"
- "What might someone who disagrees say?"
- "How does this connect to what we learned yesterday?"
- "What pattern do you notice here?"
- "What assumption might you be making?"
- "What would happen if that variable changed?"
- "What evidence supports your idea most strongly?"

Questions like these encourage students to move beyond quick answers and engage with the reasoning behind their ideas. They ask students to interpret information, use judgment, defend a position, and explain their reasoning. These processes strengthen understanding in ways that simple answer retrieval cannot. When adults regularly ask questions that invite explanation and reflection, students begin to see learning differently.

The task is no longer just finding the answer. The task is understanding why the answer makes sense. In this way, paraeducators help create classrooms where thinking remains central. You are not simply helping students produce answers. You are helping them develop the habits of mind that allow them to become thoughtful, independent learners.

Question Stems by Grade Band

Paraeducators often ask for concrete language they can use in the moment.

Questions do not need to be complex to support thinking. In fact, research consistently shows that clear, focused prompts are often more effective than lengthy explanations when supporting student learning (Rosenshine, 2012). Simple questions help students pause, organize their thinking, and engage with the task rather than waiting for the adult to provide the answer.

Different age groups benefit from different kinds of questions. Developmental research suggests that younger learners benefit most from explicit cues and strategy reminders, while older students benefit from questions that require analysis, explanation, and interpretation (Bransford, Brown, & Cocking, 2000; Hattie, 2012).

For younger students, questions that draw attention to visible strategies – such as breaking words into parts or identifying the first step of a problem – support early skill development and working memory. Cognitive science research shows that novice learners benefit from clear guidance and structured prompts as they build foundational knowledge (Sweller, Ayres, & Kalyuga, 2011).

As students grow older and develop stronger background knowledge, questions can shift toward deeper reasoning. Prompts that ask students to explain their thinking, defend a claim, or interpret evidence support higher-level learning processes such as analysis, transfer, and argumentation (Chi, 2009; Hattie, 2012).

For paraeducators, the goal is not to craft the perfect question every time. The goal is to use short prompts that keep the student thinking.

The examples below offer question stems that can be used across subjects and grade levels.

Elementary Example

Questions That Guide Visible Strategies

At the elementary level, questions should be simple, concrete, and closely tied to visible strategies. Younger students benefit from prompts that help them notice patterns, break tasks into parts, and use strategies they have already learned.

When reading: "What sound does that chunk make?" "Does that look right?" "Does that make sense?" "Try breaking the word into parts." "What clue helped you figure that out?"

When writing: "What are you trying to say here?" "Can you tell me a little more about that idea?" "Where does your sentence begin?" "What word might make that clearer?"

When solving math problems: "What's the first step?" "How many groups do you see?" "What operation might fit here?" "Can you draw a picture to show your thinking?"

Elementary students often benefit from:

- Concrete cues
- Breaking tasks into smaller pieces
- Visible reminders of strategies

Keeping prompts short and direct helps students stay focused on the task.

Middle School Example

Questions That Develop Reasoning

Middle school students are developing stronger reasoning skills while also becoming more aware of their identity as learners.

Questions at this level can begin to emphasize patterns, connections, and explanation.

When analyzing information: “What pattern do you notice?” “What changed between these examples?” “What relationship do you see here?” “What might explain that result?”

When writing: “What is your claim?” “What evidence supports that idea?” “How does that connect to the question?” “What might someone argue against that point?”

When problem solving: “What strategy did you choose?” “Why did you choose that approach?” “Where do you think the error might have happened?” “What could you adjust to try again?”

At this stage, questions should encourage:

- Ownership of ideas
- Explanation and defense
- Connections between concepts

Middle school prompts encourage ownership of ideas, explanation of reasoning, and connections between concepts.

High School Example

Questions That Build Argument and Interpretation

High school students benefit from questions that invite analysis, interpretation, and argument.

Rather than focusing only on correct answers, prompts can help students evaluate evidence and consider multiple perspectives.

When writing essays: “What is your thesis?” “What is your strongest piece of evidence?” “How does this paragraph support your argument?” “What counterargument might challenge this idea?”

When analyzing data: “What trend stands out to you?” “Which variables matter most here?” “What assumption might be influencing your interpretation?” “What does the evidence suggest, even if it doesn’t fully prove it?”

When reading complex texts: “What is the author’s purpose?” “What claim is the author making?” “Which evidence seems strongest?” “What perspective might be missing?”

Questions at the high school level help cultivate:

- Argument and justification
- Interpretation of evidence
- Transfer of ideas across contexts

High school prompts emphasize argument, interpretation of evidence, and transfer of ideas across context.

Protecting Thinking in the Moment

Many of the most important instructional moments happen quickly. A student hesitates. A question is asked. An adult responds. In those brief exchanges, the response can either replace the student's thinking or protect it.

When a student says: "I don't know." Pause before answering. Instead of filling the silence, you might ask: "What part do you know so far?" This small shift invites the student to reconnect with what they already understand.

When a student says: "Just tell me." You might respond honestly but calmly: "If I tell you, you won't build the skill." Then soften the moment by offering support: "Let's figure it out together." This keeps the student involved in the process rather than removing the challenge entirely.

When a student appears overwhelmed or anxious, the task may simply feel too large. In those moments, reducing the demand can help restore momentum. You might say: "Let's just start with the first step."

Small successes often rebuild confidence and help students reengage with the work.

Example: Elementary Reading

A student pauses while reading and asks: "What's this word?" You could say the word quickly so the sentence continues. Instead, you guide the decoding process. "What sound does the first chunk make?" The student hesitates.

You offer another prompt: "Try blending it slowly." After a moment, the student reads the word correctly. You name the strategy: "You broke the word into parts. That worked."

When the next word appears, the student attempts it independently.

Thinking remains with the learner.

Example: Middle School Lab

During a science activity, a student looks up and asks: "Is this right?" You could confirm the answer immediately. Instead, you redirect the student back to the evidence. "What evidence supports your conclusion?"

The student looks again at the data table. They compare results. They revise their explanation.

Ownership of the analysis increases because the student remains responsible for interpreting the information.

Example: High School Essay

A student stares at their outline and says: "I don't know what to argue." You could suggest a thesis and help them move forward.

Instead, you invite the student to clarify their thinking. "What position do you lean toward right now?" They offer a rough idea. You continue: "What evidence supports that position?"

The student begins identifying supporting points. Gradually, the structure of the argument takes shape.

You did not write the thesis. You protected the student's opportunity to develop it.

The Most Powerful Question

If you remember only one phrase from this chapter, let it be this: "Talk me through your thinking."

This simple question changes the direction of the moment.

Instead of focusing only on the answer, it invites the student to describe the reasoning behind their approach.

The question slows the interaction and encourages the student to explain what they noticed, what they attempted, and how they arrived at their idea.

In that moment, several important shifts occur. The focus moves:

From answer to reasoning
From panic to process
From dependence to ownership

When students explain their thinking, they organize their ideas, identify gaps in their reasoning, and deepen their understanding of the task.

The adult no longer becomes the source of the answer. Instead, the adult becomes a partner in helping the student clarify their thinking.

One of the strengths of this question is its flexibility. It works during reading, writing, mathematics, science, and discussion.

Whenever a student pauses, guesses, or asks for confirmation, the same prompt can bring the thinking back to the learner.

"Talk me through your thinking."

Sometimes the most powerful support is simply creating space for the student to explain what is already forming in their mind.

Implementation

Tracking Your Language

The words adults use during instruction influence how students approach learning. One way to reflect on your practice is to notice how often different types of responses occur during the day. For one week, pay attention to the kinds of language you use when students ask for help.

You might track moments when you:

- Provided the answer directly
- Offered a prompt or hint
- Asked a guiding question
- Asked the student to explain their thinking

The goal is not to judge your responses. The goal is simply to become aware of patterns. At the end of the week, take a moment to reflect. Ask yourself: Did my language encourage more thinking, or did it replace the student's thinking? Small shifts in language can gradually change how students approach difficulty.

When questions and prompts become more common than answers, students often begin to engage with the work differently.

One early sign of growth appears when students start working through their thinking before looking to the adult for confirmation. They reread the problem. They attempt a strategy. They explain their reasoning.

In those moments, the student is no longer waiting for the answer.

They are beginning to think their way toward it.

Closing Reflection

Paraeducators are not responsible for designing the lesson.

But you are present during some of the most important moments in learning – the moments when a student pauses, hesitates, or asks for help.

In those moments, your words shape what happens next.

A quick answer may help the assignment move forward. But a thoughtful question can help the student engage more deeply with the task.

Answers often complete the work. Questions strengthen the thinking behind it. When students are invited to explain their reasoning, try a strategy, or reflect on what they already know, they remain the ones doing the intellectual work. Over time, those moments accumulate.

Students begin to approach problems with more confidence. They attempt solutions before asking for help. They recognize that thinking – not just answers – is the goal.

You may not design the lesson. But through the language you use beside the desk, you help shape how students experience learning.

Answers may finish the assignment. Questions help build thinkers. And thinkers are what last.

Chapter 6: AI in Your Hands

Artificial intelligence is now present in many classrooms. Sometimes it appears on student devices. Sometimes it is built into writing platforms or learning tools. Sometimes teachers invite students to use it as part of the lesson. And sometimes students experiment with it quietly on their own.

Because paraeducators often work closely with students during independent work, you may be one of the first adults a student turns to when AI appears in the learning process.

A student may ask: "Can I use AI for this?" "Is this allowed?" "Can it help me figure this out?" Moments like these require clarity. Not fear. Not panic. Not overreaction. What they require is professional judgment.

AI tools can sometimes support learning by helping students organize ideas, generate examples, or review information. But those same tools can also bypass the thinking that the assignment is meant to develop.

In other words, AI can either support independence or replace thinking. The difference often depends on how it is used.

Paraeducators do not need to be technology experts to navigate these situations. But it is important to recognize that your interpretation in these moments matters.

When a student reaches for AI, your response can influence whether the tool becomes a support for thinking or a shortcut around it.

This chapter explores practical ways paraeducators can approach AI use in the classroom – so that technology supports learning while protecting the student's role as the thinker.

AI Is Not the Enemy

Artificial intelligence tools are becoming common in classrooms. Many of these tools can perform tasks that support students as they work through assignments. For example, AI tools can read text aloud, generate outlines, summarize passages, offer sentence starters, provide vocabulary definitions, translate text, model examples, and check grammar. For some students, these features can reduce barriers that make learning more difficult.

Students who experience executive functioning challenges may benefit from help organizing ideas. Students learning English may benefit from translation or vocabulary support. Students with decoding difficulties may benefit from hearing text read aloud. When used thoughtfully, these tools can increase access to learning.

But access is not the same as ownership.

A student may complete a task with AI assistance while still missing the opportunity to engage with the thinking the assignment was designed to develop. For that reason, the most important question is not simply: "Is AI allowed?" The more important question is: "What is AI doing to the thinking?"

If AI helps a student organize, clarify, or support their own thinking, it can function as a helpful tool.

If AI replaces the student's reasoning entirely, the learning opportunity disappears.

The role of the adult in the room is to help students use tools in ways that support thinking rather than replace it.

AI as Mirror vs. AI as Crutch

Artificial intelligence can play very different roles in the learning process. The key difference lies in whether the tool supports thinking or replaces it.

Understanding this distinction helps adults guide students toward responsible and productive use of AI. One helpful way to think about this difference is through two roles AI can take: a mirror or a crutch.

AI as a Mirror

When AI functions as a mirror, it reflects the student's thinking back to them. The student still generates the ideas, attempts the work, and remains responsible for the reasoning. The tool simply helps the student examine or organize what they have already produced.

In this role, AI might:

- Provide feedback on an idea
- Suggest alternative phrasing for a sentence
- Model the structure of an essay or explanation
- Ask clarifying questions about a response
- Help organize ideas the student has already generated

In these situations, the student remains the thinker.

AI does not produce the core ideas. Instead, it helps the student refine and clarify them.

Used in this way, AI can support independence and help students develop stronger thinking habits.

AI as Crutch

Artificial intelligence can also function in a very different way. Instead of supporting a student's thinking, it can begin to replace it. When AI becomes a crutch, the tool performs the core cognitive work that the assignment was designed for the student to practice.

In these moments, AI might:

- Generate the thesis for an essay
- Write an entire paragraph
- Solve the equation without explanation
- Produce the interpretation of a text or graph
- Provide the answer with little effort from the student

The assignment may still appear complete. But the thinking behind it may not belong to the student. Instead of acting as the author of the ideas, the student becomes the editor of AI-generated work.

Over time, this shift can weaken independence. Students may begin relying on the tool to do the difficult thinking rather than developing their own strategies for solving problems.

The key distinction comes down to cognitive ownership. A helpful question to ask in any situation involving AI is simple: Who is doing the heavy lifting?

If the student is generating the ideas and using AI to refine them, the tool functions as support.

If AI is producing the thinking itself, the learning opportunity begins to disappear.

When AI Supports Independence

Artificial intelligence can support learning when it reduces barriers without replacing the student's thinking. Some students face challenges that make it difficult to access the task itself. Decoding difficulties, language barriers, and executive functioning challenges can prevent students from engaging with the core ideas of a lesson.

In these situations, AI tools can sometimes help remove obstacles so that the student can focus on the thinking the assignment is meant to develop. For example, AI may be helpful when it reduces decoding barriers so comprehension can occur, translates directions so a student can access the task, helps organize scattered notes into a clearer structure, provides definitions for unfamiliar vocabulary, models the format of a response without generating the content, checks grammar after the student has written their own ideas, and summarizes a passage after the student has already read it.

In these situations, the technology removes friction from the learning process without taking ownership of the thinking. The student still remains responsible for the important cognitive work. The student still:

- Generates ideas
- Chooses strategies
- Interprets meaning
- Defends reasoning

When AI functions this way, it acts as a support tool rather than a substitute for thinking.

The goal is not to eliminate helpful tools. The goal is to ensure that the student remains the primary thinker in the learning process.

When AI Replaces Thinking

Artificial intelligence becomes problematic when it begins to replace the student's thinking rather than support it. This can happen when the tool completes the core intellectual work that the assignment was designed for the student to practice.

For example, AI may begin to replace thinking when it:

- Generates full responses for the student
- Supplies interpretations without the student first attempting the analysis
- Produces conclusions the student cannot explain
- Writes essays the student cannot defend in discussion
- Solves math problems the student cannot replicate independently
- Provides answers before the student has had a chance to struggle productively

In these situations, the assignment may appear finished, but the learning opportunity has been reduced.

One simple way to recognize this pattern is to ask a student to explain what the AI produced. If the student cannot explain the reasoning, the ideas, or the process behind the response, then the thinking likely did not belong to them.

When that happens, ownership is lost.

And when students rely on output they do not fully understand, their confidence becomes fragile.

Confidence built on borrowed answers often collapses when students are asked to explain, defend, or apply the idea on their own.

Protecting student thinking means ensuring that tools assist the process without replacing the learner's role as the thinker.

AI Thinking Protection Scale

Artificial intelligence tools can sometimes support learning and sometimes replace the thinking the assignment is designed to develop. Before helping a student use AI, it can be helpful to pause and consider how the tool is affecting the student's thinking.

One simple way to do this is to reflect on a few key questions. For each question, rate the situation on a scale from 1 (low risk to thinking) to 5 (high risk to thinking).

1. Is the student generating original ideas before using AI?
2. Can the student explain the AI output without looking at it?
3. Is AI helping organize the student's ideas, or producing the ideas itself?
4. Is AI checking grammar or formatting, or constructing the main argument?
5. Would the student likely be able to complete a similar task independently in the future?

After reflecting on these questions, consider the overall pattern. If most responses fall in the 4–5 range, the tool may be replacing the student's thinking rather than supporting it. If most responses fall in the 1–2 range, the AI tool may be helping remove barriers while the student still does the cognitive work.

This reflection does not mean AI should always be avoided. Instead, it encourages a moment of professional judgment before approving or encouraging AI use. A helpful guiding question in these situations is simple: "Who is doing the thinking here?"

If the student remains the thinker, the tool may be supporting learning. If the tool is doing the thinking, the learning opportunity may be disappearing.

The "Human Layer" Reflection

When students use artificial intelligence during learning, it is important that their own thinking remains visible. One way to protect this is by adding a simple step after AI is used – a human layer of reflection.

This step invites the student to examine the output and reconnect it to their own reasoning. Rather than accepting AI suggestions automatically, students pause to explain, revise, and evaluate what the tool produced.

You might ask questions such as:

- "Which part of this reflects your idea?"
- "What did you change or revise?"
- "Why did you choose that suggestion?"
- "Do you agree with this response?"
- "What part of this could you defend if someone questioned it?"
- "How would you explain this idea in your own words?"

Questions like these help bring the thinking back to the student. They encourage the learner to evaluate the output rather than simply accepting it. If a student cannot explain what the AI produced, it may be a sign that the thinking did not fully belong to them.

The goal is not to eliminate useful tools. The goal is to ensure that AI does not become the final voice in the learning process.

AI can function as a helpful draft partner – offering suggestions, structure, or feedback. But the student must remain the author of the ideas.

Transparency Script

When artificial intelligence tools appear in classroom work, clarity is important.

Students may experiment with AI tools in ways that feel informal or uncertain. In these moments, it is important that adults avoid encouraging AI use quietly or informally.

Transparency protects both students and educators.

Clear language helps establish expectations about how AI can be used while keeping the student responsible for the thinking. You might say:

- "If we use AI here, we need to make sure the thinking still comes from you."
- "Your teacher needs to know how this tool is being used."
- "AI can help organize your ideas, but the ideas still need to come from you."
- "Let's use AI to check grammar after you finish writing."
- "Can you explain this idea without looking at the screen?"

Statements like these help students understand that technology is a tool for support, not a substitute for their thinking. They also reinforce that AI use should remain visible to the teacher and aligned with classroom expectations.

Transparency helps build a culture of ethical technology use.

When AI use is discussed openly, students learn how to use tools responsibly. When technology is used secretly or without guidance, confusion and risk increase.

Clear communication keeps the focus where it belongs – on learning and student thinking.

Ethical Guardrails for Paras

Artificial intelligence tools are becoming more common in classrooms, but their use requires thoughtful boundaries. Paraeducators play an important role in maintaining those boundaries while still supporting students with dignity and professionalism. Clear guardrails help ensure that technology supports learning rather than undermining it. Consider the following principles when AI tools appear during student work:

1. Follow the teacher's expectations first. Classroom policies about AI use should always guide your response.

2. Avoid introducing new AI tools without alignment. If a tool has not been discussed with the teacher, pause before encouraging its use.

3. Do not complete student work using AI. AI should never be used by an adult to produce answers or assignments for a student.

4. Invite explanation after AI use. Ask the student to explain the ideas, reasoning, or revisions connected to the AI output.

5. Prioritize long-term skill development. Finishing an assignment quickly matters less than helping students develop the skills the assignment is meant to build.

6. Protect academic integrity. Encourage students to represent their own thinking honestly.

7. Notice patterns of misuse. If AI appears to be replacing student thinking consistently, communicate those observations with the teacher.

8. Maintain dignity in every interaction. When concerns arise, approach the conversation with curiosity and respect rather than accusation.

The role of the paraeducator is not to act as a technology police officer. Instead, the role is to protect the learning process. AI may be present in the classroom, but the student's thinking must remain at the center of the work.

Elementary Example

Reading Support

A student opens a reading assignment and activates an AI tool that reads the passage aloud.

For this student, decoding printed text is challenging. The tool allows them to hear the words so they can access the meaning of the passage.

In this moment, the AI tool is helping remove a barrier. It is supporting access rather than replacing the task.

After the student listens to the paragraph, you pause and ask: "What was the main idea of that section?"

The student thinks for a moment and explains the central idea in their own words. You follow with another prompt: "What detail helped you figure that out?"

The student points to a sentence in the text. Now the student is interacting with the meaning of the passage, not just listening to the words.

If the student cannot explain what they heard, it may signal that listening replaced engagement with the text.

In that case, you might guide them back with a smaller step: "Let's listen to that sentence again. What do you think it is mostly about?"

AI supported decoding.

But comprehension – the thinking part – still belongs to the student.

Middle School Example

Lab Report Draft

During a science activity, a student copies the lab question into an AI tool and receives a fully written conclusion.

They show the response to you and ask: “Is this right?” You could quickly approve the answer so the assignment moves forward.

Instead, you shift the focus back to the student’s reasoning. “Can you explain why this conclusion makes sense?”

The student hesitates and looks back at the screen. This hesitation is an important signal. The explanation may not yet belong to the student.

You redirect their attention to the original data. “Let’s go back to your results. What pattern did you notice across the trials?”

The student begins scanning the data table again.

Slowly they describe what changed between the experiments.

You follow up: “How does that pattern connect to the conclusion?”

Now the student begins revising the explanation so it reflects their own observations.

In this moment, AI had replaced the analysis. Your questions returned the student to the role of the thinker.

Ownership of the explanation begins to grow again.

High School Example

Essay Planning

A student working on an argumentative essay uses AI to generate a possible thesis statement. They show you the sentence and ask: "Does this sound good?"

Instead of evaluating the thesis immediately, you ask a different question: "Do you agree with this claim?"

The student shrugs slightly.

This response suggests that the idea may not fully belong to them yet. You continue: "What position do you actually want to argue?" The student pauses and begins explaining their perspective on the topic.

You listen, then respond: "That sounds like a strong argument. How could we turn that idea into a thesis?"

Together, the student begins adjusting the AI-generated sentence so that it reflects their own thinking.

You might ask: "What evidence would support that claim?" The student lists several examples and begins shaping the outline for the essay.

In this moment, AI becomes a mirror rather than a crutch. It provided a starting point, but the student revised and reshaped the idea so that the argument reflects their reasoning.

The student remains the author.

Ownership of the work – and the thinking behind it – stays with them.

The Emotional Temptation

Artificial intelligence can make schoolwork feel faster. A student enters a prompt. A response appears instantly. The assignment moves forward.

In these moments, students often feel relief. The confusion disappears. The pressure eases. The work feels easier.

Paraeducators may also feel helpful when the task moves along quickly. But speed is not the same as growth.

An assignment can be completed without the student developing the skill the work was meant to build.

Completion is visible. Competence develops more slowly.

For that reason, it can be helpful to pause and reflect in moments when AI appears to solve the problem quickly. Ask yourself: Am I protecting the student's independence – or enabling avoidance of the thinking?

Shortcuts can feel efficient in the moment.

But long-term skill development often requires students to experience challenge, attempt strategies, and work through uncertainty.

Those moments of effort may feel slower.

Yet they are the moments when real learning begins to take shape.

AI and Confidence

Artificial intelligence can make answers appear quickly. But when students rely on a tool too early in the learning process, it can quietly change how they see their own ability.

If AI becomes the first step rather than a later support, several patterns may begin to emerge. Students may begin to feel less confident attempting difficult work on their own. Over time self-efficacy decreases, and risk tolerance drops.

When students repeatedly turn to a tool before attempting the task themselves, they miss opportunities to experience the small moments of success that build confidence.

Confidence grows through effort.

It develops when students attempt a challenge, adjust their strategy, and eventually reach understanding.

Those experiences create an important internal message: "I figured this out."

When a tool consistently provides the answer instead, the message becomes different: "The tool figured it out – not me."

Protecting student thinking means ensuring that technology supports learning after effort, not in place of it.

When students remain the ones solving problems and explaining ideas, confidence has a chance to grow alongside skill.

IMPLEMENTATION

Artificial intelligence will appear in classroom work more often. Students may experiment with tools during writing, research, or problem solving. When that happens, the goal is not to react quickly or shut the tool down immediately.

The goal is to guide the moment so the student's thinking remains central.

This week, try focusing on just one interaction involving AI. When a student uses an AI tool, take a moment to structure the conversation around thinking. You might:

1. Ask one Human Layer question that invites the student to reflect on the AI output.
2. Require one explanation in the student's own words about what the response means.
3. Ensure one part of the work remains fully student-generated, such as an idea, example, or explanation.
4. Communicate openly with the teacher if the use of AI raises questions about expectations or boundaries.

After the interaction, take a moment to observe what happened. Ask yourself: Did the student's thinking increase during the conversation? Did defensiveness decrease when curiosity guided the interaction? Did the student show stronger ownership of the ideas?

Small conversations like this can shift how students approach AI.

Instead of relying on the tool to produce answers, they begin to see it as something that can support their thinking while the responsibility for learning remains their own.

Closing Reflection

Artificial intelligence is becoming a permanent presence in many classrooms. Students will encounter these tools in writing platforms, research tools, and learning applications. Over time, their use will likely become even more common. But as technology evolves, one principle must remain clear.

Thinking cannot disappear. Paraeducators often stand at an important intersection during moments of learning. You are present when a student turns to a tool. You are present when confusion appears. You are present when a decision must be made about how to move forward.

In those moments, you stand between several possibilities:

Between tool and student.
Between shortcut and stamina.
Between completion and comprehension.

Artificial intelligence can accelerate answers. But it cannot build judgment.

Judgment grows when students interpret information, weigh evidence, and explain their reasoning.

Those habits of thinking remain deeply human work.

When guided thoughtfully, AI can still play a useful role.

In the hands of a careful paraeducator, the tool becomes something that reflects thinking rather than replaces it.

AI can act as a mirror – helping students refine and organize their ideas. But the thinking itself must remain with the learner.

PART IV — WORKING IN ALIGNMENT

Chapter 7: Supporting the Teacher's Design

Classrooms are designed environments. Every lesson a teacher plans includes decisions about pacing, questioning, difficulty, and the type of thinking students are expected to do.

Paraeducators step into that environment already in motion. You enter the room. The lesson is already in motion. The objective is posted. The activity begins. You move beside students and begin supporting their work.

At first, the role may appear simple: help students complete the task. But an important tension exists beneath the surface. If the instructional design behind the lesson is not fully understood, support can unintentionally shift the learning in a different direction. Not intentionally. But powerfully.

A hint given too quickly may remove the thinking the teacher wanted students to attempt. An answer offered too early may replace the struggle the lesson was designed to create. A scaffold added at the wrong moment may change the level of challenge the teacher carefully planned.

In these moments, the adult beside the desk is influencing the learning architecture of the lesson. Supporting students effectively therefore requires more than helping individuals finish their work. It requires working in alignment with the teacher's instructional design.

This chapter explores how paraeducators can read the structure of a lesson, recognize the thinking it is meant to produce, and provide support that strengthens the teacher's design rather than unintentionally replacing it.

Because when support and instruction move in the same direction, something powerful happens. Students receive help. But their thinking remains intact.

What Is Instructional Architecture?

Instructional architecture is the invisible structure beneath a lesson.

It determines where cognitive tension is introduced, when struggle is expected, how confusion is sequenced, when release occurs, where explanation is required, and how feedback is timed.

Teachers do not simply assign work. They design thinking experiences.

As explained in *AI and the Art of Teaching*, strong lessons are built to protect the student's role as the thinker. The structure of a lesson often determines whether students experience genuine reasoning or simply complete tasks.

Sometimes that design intentionally includes front-loaded thinking before explanation, delayed explanation, silent think time, productive struggle, peer discourse, and defense of reasoning.

These elements are not accidents. They are part of the lesson's design.

If support arrives too quickly, the architecture changes. A hint may replace thinking. An answer may remove the struggle the lesson was designed to create. A scaffold may appear before the student has attempted the work.

In those moments, the task becomes easier. But something else may quietly disappear.

You may reduce frustration. But you may also reduce growth.

Why Teachers Design Cognitive Tension

Cognitive tension is the moment in a lesson when thinking begins to stretch.

It appears when students are unsure, the answer is not obvious, confusion rises slightly, and thinking becomes necessary. These moments can feel uncomfortable. But they are not signs of poor teaching. They are often intentionally designed.

Research on learning shows that deeper understanding develops when students grapple with ideas before receiving direct explanation. This process is productive, where initial struggle prepares the mind to understand later instruction more deeply (Kapur, 2008).

Without cognitive tension there is no reasoning, here is no transfer, there is no endurance, there is no deep learning. Because of this, teachers often design lessons around a sequence like this:

Tension → Struggle → Strategy → Release

Students encounter difficulty. They attempt to make sense of the problem. Strategies begin to form. Then explanation or clarification brings resolution.

If support removes the tension too early, the cycle changes.

Students may reach the answer faster. But the thinking that builds understanding may never occur.

When cognitive tension disappears too quickly, the learning cycle collapses.

Why Struggle Is Intentional

Struggle is often the moment when real learning begins.

It signals that the brain is working, new connections are forming, and understanding is being constructed. When students encounter a problem they cannot solve immediately, the mind begins searching for patterns, strategies, and explanations. That mental effort is part of how learning develops.

To students, however, struggle can feel very different. Students can interpret struggle as: "I'm bad at this." "I can't do this." "I should already understand."

When answers do not appear quickly, students may assume something is wrong with them rather than recognizing that the difficulty is part of the learning process.

Teachers interpret struggle differently. They recognize it as a normal and necessary stage of skill development. They understand that moments of uncertainty often mean students are actively working through a new idea. From the teacher's perspective, struggle is often evidence that the lesson is doing what it was designed to do. Paraeducators must learn to adopt the same interpretation.

Struggle does not mean something is going wrong. Often it means learning is happening. When students remain engaged during difficulty, they are strengthening the very abilities that lead to independence.

What matters most is how the struggle is supported. Struggle without humiliation builds strength. Struggle with rescue builds dependence. Struggle with dignity builds confidence.

When students are allowed to wrestle with ideas while feeling safe and supported, they begin to discover something powerful: they are capable of thinking through difficulty. And that discovery is one of the foundations of lasting confidence.

Why Rescue Can Undermine Rigor

Rescue often feels helpful. You reduce discomfort. You speed up completion. You ease visible frustration.

In the moment, it can feel like the right thing to do. When a student looks stuck or discouraged, stepping in quickly may seem like the most supportive response.

But rescue can also change the learning process. When answers arrive before students attempt the work themselves, an important part of the learning experience can disappear.

Rescue may prevent productive struggle, short-circuit reasoning, lower cognitive demand, create reliance on adult proximity, and undermine the teacher's instructional goals.

Sometimes what looks like silence from the teacher is intentional.

The teacher may be waiting. Waiting for students to attempt. Waiting for thinking to emerge. Waiting for the class to wrestle with the idea.

Those pauses are often part of the lesson design. They give students time to process, test ideas, and discover what they can do on their own.

If support appears during that pause, the design of the lesson changes.

A moment meant for thinking becomes a moment of assistance. A moment meant for reasoning becomes a moment of explanation. Support is most powerful when it moves in the same direction as the lesson.

When support aligns with the teacher's design, students receive help without losing their role as thinkers.

Support is most powerful when it aligns with the instructional architecture.

Teacher–Para Alignment Conversation Template

Paraeducators work inside lessons they did not design. Because of this, clarity matters. When teachers and paraeducators share a clear understanding of the lesson's purpose, support becomes more precise and more effective.

Brief alignment conversations can prevent confusion later in the lesson. Use this template during planning time, before a lesson begins, or during short check-ins throughout the week. Ask the teacher:

1. What is the main cognitive goal of this lesson?
2. Where should students struggle before receiving help?
3. When should I step in – and when should I wait?
4. What level of prompting is appropriate?
5. What language are you using so I can mirror it with students?
6. What should independence look like during this activity?
7. Are there specific students who need prompt fading?
8. Are there phrases or supports I should avoid using?

These conversations do not need to be long. Even a brief moment of alignment can clarify expectations and prevent accidental interference with the lesson's design.

Alignment reduces confusion. It increases coherence. Students experience unified adult messaging.

When adults work from the same plan, support strengthens the learning instead of unintentionally redirecting it.

“Before I Help” Clarifying Questions

Paraeducators often notice confusion before anyone else in the room.

Because you are close to students, you may see hesitation, frustration, or uncertainty before the teacher does. In those moments, the instinct to help can appear quickly. Helping feels supportive, and most adults want to reduce discomfort when students begin to struggle. But pausing briefly before stepping in can protect the learning design of the lesson.

Not every moment of confusion needs immediate assistance. Sometimes the teacher has intentionally created space for students to think, test ideas, and work through uncertainty. Before offering help, take a brief mental pause and ask yourself:

1. Is this confusion not

2. Has the teacher already introduced the strategy students should try?

3. Is silence part of the lesson’s design?

4. Is this student in productive struggle – or moving toward shutdown?

5. Will stepping in now strengthen independence or weaken it?

6. Am I helping because the student feels uncomfortable – or because they are truly overwhelmed?

These questions take only a few seconds to consider. But that brief pause can make an important difference in how the learning unfolds. They help prevent premature rescue. They protect productive struggle. They honor the teacher’s instructional architecture. And most importantly, they help keep the student in the role of the thinker, where real learning takes place.

Design Lens Observation Guide

Paraeducators often support lessons while they are already in progress. Taking a moment to observe the structure of the lesson can help you understand what kind of thinking the teacher is trying to create. Instead of focusing only on the assignment students are completing, try to notice the sequence of the lesson itself.

This observation helps reveal the instructional architecture behind the activity. When observing a lesson, look for moments such as:

☐ Where cognitive tension increases
☐ When the teacher delays answering a question
☐ When peer discussion is encouraged
☐ When explanation is required
☐ When feedback is temporarily withheld
☐ When struggle is normalized
☐ When release or clarity is provided

These signals help reveal how the lesson is designed to unfold. Then ask yourself a simple question: What is the sequence of the lesson?

Many strong lessons follow a pattern such as:

Confusion → Strategy → Practice → Reflection

Students first encounter uncertainty. They attempt strategies to make sense of the problem. They practice applying the idea. They reflect on what they learned.

When you recognize this flow, your support can mirror it.

Support should reinforce the sequence of the lesson. Not override it.

Elementary Example

Math Word Problem

During a math lesson, students are working through a word problem. Before anyone begins solving, the teacher gives a direction to the class: "Don't solve yet. Talk about possible strategies."

Students begin discussing with partners. One student looks unsure. They glance at you and ask quietly, "What do we do?"

You feel the instinct to explain the strategy so the student can move forward. But you pause.

The teacher has not asked students to solve the problem yet. The goal of this moment is strategic thinking, not answers.

Instead of explaining the solution, you redirect the thinking. You ask: "What strategies did your teacher mention earlier?"

The student begins recalling ideas from the lesson. They turn back to their partner and discuss possibilities.

You did not solve the problem. You kept the thinking inside the lesson design. You supported the strategy conversation the teacher intended.

The instructional architecture remains intact.

Middle School Example

Socratic Discussion

During a class discussion, the teacher asks a question: “What evidence supports that claim?”

Students review the text and begin responding. One student starts to speak but hesitates. They look unsure and glance toward you.

You lean slightly closer to help. Then you pause. Is the hesitation confusion – or part of thinking?

In Socratic discussions, silence often signals that students are searching for evidence. The teacher is giving space for that thinking to occur.

Instead of supplying the evidence, you offer a quiet prompt. You whisper: “What part of the text connects to that idea?”

The student looks back at the page and scans the paragraph.

After a moment, they point to a sentence and share it with the class.

You did not provide the answer. You prompted the student to locate the evidence themselves.

The reasoning process remained the student’s work. The discussion architecture stayed intact.

High School Example

Essay Draft Workshop

During a writing workshop, students are reviewing each other's essays. Before the activity begins, the teacher gives a clear instruction to the class: "Don't rewrite your partner's work. Ask questions that help them revise."

Students exchange drafts and begin reading. One student turns to you and says: "Can you fix this paragraph?"

You could quickly rewrite the sentence or suggest a stronger version. But that would take ownership of the writing.

Instead, you mirror the teacher's goal. You respond with a question: "What is your main argument in this paragraph?"

The student rereads their writing and explains their idea.

You follow with another prompt: "What evidence supports that point?"

The student begins revising the paragraph themselves.

You did not edit the writing for them. You supported the revision process while keeping the student as the author.

The instructional design of the workshop remained intact.

Alignment Protects Students

Students pay close attention to how adults respond during learning.

Even when they are focused on the task, they notice how different adults react to confusion, mistakes, and struggle. Those reactions help students decide what learning is supposed to look like.

When the teacher and paraeducator respond in different ways, students receive mixed signals about what is expected. One adult rescues. One adult expects struggle.

Students may not know which expectation to follow. Should they attempt the work independently? Or should they wait for help to arrive?

When this happens, confusion increases. Dependence increases. Frustration increases.

Over time, the student may begin to rely on the adult who removes the difficulty rather than the adult who is encouraging thinking. The learning environment becomes less predictable, and the student's confidence in their own ability to try may weaken.

Alignment changes this experience. When the teacher and paraeducator operate from the same expectations, students encounter a consistent message about learning. Struggle is expected. Thinking is required. Support is available – but it protects independence.

Students begin to understand that difficulty is part of the process rather than a signal to stop.

Alignment creates consistency. And consistency helps students feel safe enough to keep trying.

Consistency builds confidence.

The Emotional Reality

Paraeducators often work in moments of uncertainty.

You are close enough to see when a student becomes confused, frustrated, or discouraged. At the same time, you want to respect the teacher's instructional plan and the structure of the lesson.

This can create a quiet tension. Paras often feel caught between supporting the student, respecting the teacher, and managing visible struggle in the moment.

When a student looks stuck, the instinct to help appears quickly. Waiting can feel uncomfortable, especially when frustration is visible.

Clarity reduces that tension.

When you understand the instructional architecture of the lesson, your decisions become easier and more intentional. You begin to see where struggle is expected and where support should appear.

When you understand the architecture, you trust struggle more, intervene more precisely, fade support more intentionally, and protect the rigor of the lesson.

Your role becomes less reactive and more strategic.

You are not "less helpful" when you wait. You are helping in a different way. You are protecting the thinking the lesson was designed to create. You are being more strategic.

IMPLEMENTATION

One Alignment Conversation

Alignment does not require long meetings or complicated planning. Often, a brief conversation can create enough clarity to improve how support appears during the lesson.

Start with a simple step. Schedule one 10-minute alignment conversation with the teacher. Use the alignment conversation template to clarify the goals of an upcoming lesson. Focus on understanding where students are expected to think independently and where support may be appropriate.

Then observe one lesson using a design lens. Instead of focusing only on student behavior, watch the structure of the lesson itself. Notice moments such as where cognitive tension increases, where silence appears intentional, and where feedback is delayed. These signals often reveal how the teacher designed the learning experience.

For the next week, consciously match your support to the teacher's instructional architecture. As you do, pay attention to what changes. Track questions such as:

- Did premature rescue decrease?
- Did student independence increase?
- Did confusion resolve more organically through discussion or strategy?
- Did you feel more confident allowing students time to think before stepping in?

These small shifts can change how students experience learning. Alignment strengthens the lesson. It strengthens the partnership between adults. And most importantly, it strengthens student thinking.

Closing Reflection

Paraeducators do not design the lesson, but you influence how the lesson unfolds. The support you provide beside the desk can either reinforce the teacher's instructional design or unintentionally change it. Small decisions – when to wait, when to prompt, when to step in – shape how students experience the learning process.

Instructional architecture is often invisible. Yet it quietly guides how thinking develops during the lesson. It determines when students wrestle with ideas, when discussion deepens understanding, and when explanation brings clarity.

Because this structure is carefully planned, it can also be fragile. Rescue can interrupt it. Alignment protects it.

When you understand why cognitive tension exists, your role begins to shift. You stop trying to eliminate the difficulty immediately. You start recognizing the purpose behind it.

Instead of removing the struggle, you support students as they move through it. You learn to trust the process the teacher has designed. And when the teacher and paraeducator move in the same direction, something important happens. Students experience clarity. Students experience consistency. Students experience growth.

They begin to understand that struggle is part of learning and that the adults around them share the same expectations. Support becomes more than assistance. It becomes partnership.

That is professional alignment. And that is powerful support.

Chapter 8: Special Education & Inclusion Context

In inclusive classrooms, paraeducators often work closest to students with Individual Education Plans (IEPs). You are beside them during reading, math, labs, group work, transitions, during moments when the work becomes confusing, and moments when overwhelm appears. In many of those moments, you are the adult beside the students. And that proximity carries weight. Because how support is provided in these moments can shape how the student experiences school.

For many students with disabilities, the goal of support is sometimes misunderstood. Support can easily drift toward making the task easier, completing work more quickly, or preventing visible struggle. But special education support is not designed to remove challenge. It is designed to increase access.

Access means helping the student engage with the same learning goals as their peers. It means adjusting supports, prompts, tools, and scaffolds so the student can participate meaningfully in the learning process. The expectations remain. The thinking remains. The student remains the learner.

This distinction matters.

When support focuses only on making work easier, expectations can quietly drop and independence may weaken. But when support focuses on access, students gain the opportunity to attempt, struggle, revise, and grow within the learning environment.

This chapter explores how paraeducators can support students with disabilities in ways that increase access, protect dignity, and preserve the intellectual expectations of the classroom.

Because inclusion is not simply about being present in the room. It is about being supported in the work of learning.

IEP Goals are the North Star

Every student with an IEP has identified goals. Those goals are not random. They are carefully developed to address the student's specific learning needs and to guide progress over time. IEP goals reflect skill gaps, developmental needs, measurable growth targets, and specific instructions.

These goals act as a compass for the student's support. They help teachers, specialists, and paraeducators understand what the student is working toward and how progress should be measured.

When support aligns with these goals, every interaction can contribute to the student's development. But when support happens without reference to the IEP, the focus can quietly shift.

Help may begin to prioritize task completion rather than skill growth. This is why a simple question can be powerful. Ask yourself: Am I helping this student finish – or helping them progress toward their goal? Consider a few examples.

If a student has a reading fluency goal and an adult reads every passage aloud without gradually fading support, the student may complete the assignment – but their fluency practice disappears.

If a student has a writing organization goal and an adult structures every paragraph for them, the student may submit a strong paper – but the independence needed to organize ideas may never develop.

In both cases, the work gets done. But the skill the student is meant to strengthen receives fewer opportunities to grow. IEP alignment helps prevent this drift. When support consistently connects back to the student's goals, each interaction reinforces the learning the student most needs to develop.

IEP alignment protects growth. Completion alone does not.

Supporting Accommodations Without Lowering Standards

Accommodations exist to increase access. They help students participate in learning when barriers would otherwise prevent them from engaging fully with the lesson. But accommodations are sometimes misunderstood.

They are not designed to lower expectations. They are designed to remove obstacles so students can reach the same learning goals.

Under the Individuals with Disabilities Education Act (IDEA), students with disabilities are entitled to supports and services that allow them to access the general education curriculum and make progress toward their IEP goals (Individuals with Disabilities Education Act, 2004). These supports often include accommodations.

Accommodations are adjustments to how learning is accessed or demonstrated. They may involve changes to timing, presentation, response format, or environment so that a student can participate meaningfully in the same academic expectations as their peers.

In other words, accommodations change how students access learning, not what they are capable of learning.

Consider a few common examples:

- Extended time does not equal simplified content
- Read-aloud support does not equal removal of comprehension
- Graphic organizers does not equal elimination of reasoning
- Sentence starters does not equal elimination of argument

Each of these supports provides structure that helps students engage with the task.

But the thinking remains.

Students are still expected to analyze, explain, reason, and demonstrate understanding.

The academic goal does not change. Only the pathway to that goal becomes more accessible.

When expectations are lowered instead of access being increased, students receive a different message. Lowering standards unintentionally communicates: "You are not expected to reach this level." Over time, that message can weaken confidence and reduce opportunities for growth.

Access communicates something very different. "We are removing barriers so you can succeed." That message preserves both dignity and expectation.

And when students experience both support and high expectations, they are far more likely to develop independence, confidence, and progress toward their goals.

It is also important to distinguish accommodations from modifications. Accommodations change how a student accesses learning or demonstrates understanding, while the learning expectations remain the same. Modifications, in contrast, change what a student is expected to learn by adjusting the level or scope of the academic content. For example, extended time or a graphic organizer are accommodations because the learning goal stays the same. Reducing the number of standards addressed or simplifying the learning target would be considered a modification. This distinction matters because accommodations preserve access to grade-level expectations, while modifications alter the academic expectations themselves (Hall, Meyer, & Rose, 2012).

Dignity-First Modifications

Inclusion is not only academic. It is also relational.

Students do not experience support only through the assignments they complete. They also experience it through how that support appears in front of peers, how adults speak to them, and how much ownership they retain during learning. Because of this, support must protect dignity.

Dignity-first support means paying attention not only to what help is provided, but how that help is delivered. It often includes choices such as avoiding public spotlight, avoiding visible adult takeover, protecting peer perception, preserving student ownership, and using a calm, neutral tone.

These decisions may seem small, but students notice them quickly.

Students are acutely aware of difference. They observe who receives help, how often adults step in, and how visible that support becomes during class activities. If support feels constant, public, or overly controlling, it can unintentionally reinforce identity labels. The student may begin to feel defined by the support rather than by their learning.

Dignity-first design asks a different set of questions. Can this support be discreet? Can the assistance happen in a way that protects the student's role as the learner? Can independence still be signaled even while support is present? Can the student retain visible ownership of the work?

When support protects dignity, students are more willing to attempt difficult work. They feel safer taking risks and participating alongside their peers.

Inclusion should expand opportunity, not reduce identity.

Inclusion must preserve identity – not diminish it.

Independence vs. Over-Support in 1:1 Settings

One-to-one support can be extremely valuable for students. It provides consistency, immediate guidance, and an adult who can respond quickly when confusion or frustration appears. But 1:1 support also carries unique risks.

When an adult is always present beside a student, the learning dynamic can shift in ways that are not always intended. Over time, certain patterns may begin to appear. Students may begin to look to you first before attempting the work. Peers may defer to you instead of interacting directly with the student. Teachers may begin relying on the paraeducator to manage the student's engagement. And the student's independence may quietly stall.

None of these outcomes are intentional. They often emerge simply because proximity is constant. This is why 1:1 support still needs to follow the Support Spectrum used throughout the classroom:

Do for → Do with → Prompt → Question → Observe

The goal of support is always movement along this spectrum. At times, students may temporarily need higher levels of assistance. But over time, the support should gradually shift toward prompting, questioning, and observation.

Even in intensive support settings, upward movement can occur. Students need opportunities to attempt, respond, and think independently. If 1:1 support remains at Level 1 (Do for) or Level 2 (Do with) indefinitely, growth becomes limited. Assignments may be completed, but independence may not develop.

IEP support is designed to increase access and skill over time. Its purpose is not to replace the students' thinking. IEP support should increase independence – not replace it.

Accommodation Integrity Checklist

Accommodations are designed to support access. But the way they are implemented can influence whether they strengthen learning or unintentionally reduce it.

When accommodations are used thoughtfully, they help students engage with the same learning goals as their peers. When they are used automatically or without reflection, they may drift toward simplifying the task rather than supporting access.

A brief pause before implementing an accommodation can help maintain that balance. Before providing support, ask yourself:

☐ Is this aligned with the student's IEP?
☐ Does this provide access without reducing rigor?
☐ Is the cognitive demand of the task still present?
☐ Is the student generating thinking independently?
☐ Am I fading this support over time?
☐ Is this support discreet and dignity-preserving?
☐ Does this move the student toward long-term independence?

These questions help ensure that support remains aligned with the student's learning goals. They shift the focus from simply helping the student finish the task to helping the student develop the skills the IEP is designed to strengthen. If the answer to several of these questions is no, it may be worth pausing and reconsidering how the accommodation is being used.

The goal is not to remove support. The goal is to ensure the support strengthens learning.

Accommodation should not equal simplification. It should equal access.

Skill vs. Task Confusion Diagnostic

Sometimes paraeducators support the task when the real need is skill development. A student may appear stuck, and the immediate instinct is to help them finish the assignment. But finishing the task does not always address the underlying challenge the student is experiencing.

Before stepping in, it can be helpful to pause and ask a simple question: Is the student struggling because of a skill gap or because of the complexity of the task? These two situations require different types of support.

If the challenge is skill-based, the student needs opportunities to practice and strengthen that skill. If the challenge is task-based, the student may need help organizing the work or breaking it into manageable steps.

Consider an example.

A student is unable to complete an essay. The difficulty may appear to be the assignment itself. But a closer look may reveal something more specific. Ask yourself: Is the student struggling because of a lack of writing structure skills, a vocabulary gap that makes expression difficulty, an executive functioning challenge that makes planning difficult, or simply a lack of effort or engagement with the task? Understanding the source of the difficulty changes how support should appear.

If the challenge is skill-based, support the development of the skill. Provide modeling, guided practice, and opportunities for the student to attempt the work themselves. If the challenge is task-based, support the organization of the task. Break the assignment into steps, clarify expectations, or help the student sequence their work. But if an adult completes the task for the student without addressing the underlying skill, growth is unlikely to occur. The assignment may be finished. But the student's ability to complete the task independently may not improve.

Skill confusion often leads to over-support. Careful diagnosis protects independence.

Independence Planning Template

Independence rarely appears by accident. It develops when support is intentionally designed to decrease over time. Without a clear plan, adult proximity can quietly become permanent.

This template helps paraeducators and teachers think intentionally about how support can gradually shift toward independence. Start by choosing one student you regularly support. Then work through the following steps.

1. Current Support Level. Identify where support currently falls on the Support Spectrum: Do for → Do with → Prompt → Question → Observe

Understanding the current level helps clarify where growth should occur.

2. IEP Goal. What specific skill is the student working to develop? This goal should guide how support is provided. The purpose of support is not only to complete tasks, but to strengthen the skill identified in the student's IEP.

3. Current Barrier. What is currently preventing the student from working more independently? Consider possibilities such as skill gaps, task complexity, executive functioning challenges, or learned reliance on adult prompts.

4. Fading Plan. What element of support can be gradually reduced over the next four weeks? Examples might include reducing verbal prompts, increasing wait time before assistance, or encouraging the student to attempt the task before asking for help. The goal is not to remove support suddenly, but to fade it intentionally.

5. Independence Marker. How will you recognize that independence is increasing? Clear markers help adults notice growth as it happens.

Examples include:

- Initiates work without a prompt
- Asks fewer clarifying questions
- Uses a previously taught strategy independently
- Attempts the task before looking to the adult for help

These markers help adults notice when independence is beginning to grow.

Instead of relying only on general impressions, they provide observable signs that the student is beginning to take greater ownership of the work.

You may begin to notice that the student starts assignments more quickly. They may attempt strategies before asking for assistance. They may pause to think before looking toward the adult beside them.

These small shifts matter. They signal that the student is beginning to rely more on their own thinking and less on immediate support.

Independence should be visible and measurable.

When growth is defined clearly, adults can recognize progress and intentionally adjust support levels. Otherwise, support can remain constant even when students are capable of doing more on their own.

Without clear indicators of independence, it becomes easy for adult proximity to continue out of habit rather than necessity.

Independence must be measurable. Otherwise, proximity becomes permanent.

Elementary Example

Supporting a Reading Accommodation

During a reading lesson, a student with an IEP receives read-aloud support. The class begins reading a short passage together.

The student looks toward you, expecting the entire section to be read aloud. It may feel natural to begin reading immediately so the student can move through the assignment quickly. But the student's IEP goal focuses on improving decoding fluency. If the adult reads every word automatically, the student has fewer opportunities to practice that skill.

Instead of reading the entire passage, you adjust the support. You ask the student to attempt the first paragraph independently.

When the student encounters a difficult word, you support only the complex vocabulary. For familiar sections, you gradually fade the read-aloud support.

The student continues reading with increasing independence.

The accommodation still provides access. But the responsibility for reading slowly shifts back to the student. Support aligns with the goal.

Middle School Example

Supporting Note-Taking Without Taking Over

During a middle school lecture, students are recording notes from a science lesson. One student with an IEP struggles with organizing information on the page. They look at their notebook and pause.

It may be tempting to write the notes for them so they can keep up with the class. But doing so would remove the opportunity to practice the skill they are developing. Instead, you support the structure of the task.

You provide a simple note-taking template that organizes the key sections of the lesson. You model one section of the notes together.

Then you ask the student to complete the next section independently.

As the student becomes more comfortable with the structure, you gradually reduce the scaffolding. The student begins organizing the notes with less assistance.

The accommodation supports executive functioning. But it does not eliminate cognitive engagement.

The student remains responsible for the thinking.

High School Example

Supporting Extended Time in Writing

During an essay assignment, a student receives extended time as an accommodation. Because the student writes more slowly, the work may feel overwhelming at first.

It may seem easier to shorten the assignment or reduce the expectations. But extended time is not designed to lower the standard. Instead, it changes how the student accesses the task.

Rather than reducing the essay requirement, you help organize the process. You break the assignment into structured checkpoints. First, the student completes a planning outline. Next, they draft the introduction and thesis. Then they expand the body paragraphs.

You provide planning scaffolds that help the student organize their ideas. But the student must still Generates and develop the argument independently.

After the draft is complete, feedback focuses on revision and clarity.

The academic standard remains intact. The pathway to reach that standard becomes more accessible. Access is adjusted. The expectation remains.

Protecting Standards in Inclusion

Inclusion is not the removal of challenge. It is the redesign of access.

Students receiving special education support are still members of the same intellectual community as their peers. They deserve opportunities to think, reason, solve problems, and experience the satisfaction that comes from mastering difficult work.

Sometimes adults misunderstand inclusion as reducing academic expectations, so students feel successful more quickly. But when expectations are lowered too far, something important disappears. Students may complete tasks, but they lose the opportunity to build the very skills their education is meant to develop.

True inclusion protects both access and rigor.

In inclusive classrooms, the academic standard remains constant. What changes is the pathway that allows the student to reach it. Inclusion does not represent reduced challenge, lowered expectations, or adults completing the intellectual work on behalf of the learner. Research on inclusive instruction consistently emphasizes that rigorous expectations can coexist with differentiated access to learning (Tomlinson, 2014).

Instead, inclusion involves intentionally designed supports that allow students to engage with the same learning goals in accessible ways. These supports include adjusted access to the task, strategic scaffolding that guides thinking, structured supports that reduce cognitive overwhelm, gradual release of responsibility as competence increases, and the intentional development of student independence (Fisher & Frey, 2014; Tomlinson, 2014).

This distinction matters.

When challenge disappears, growth disappears with it. But when support is designed well, students experience something powerful: they discover they are capable of more than they expected.

Paraeducators play an important role in protecting this balance.

You are often the adult closest to the student during the moments when the work becomes difficult. In those moments, it can be tempting to simplify the task, provide the answer, or complete part of the work so the student can move forward quickly.

But the goal of support is not simply completion. The goal is development.

Support should help students move toward the standard, not away from it. Sometimes that means slowing the process so the student can think. Sometimes it means providing a scaffold that organizes the work. Sometimes it means asking a question that redirects the student's reasoning. These choices protect the student's role as the thinker.

Inclusion works best when support increases access without removing challenge.

When students receive this kind of support, they experience something different from rescue. They experience capability. They attempt more. They persist longer. They begin to believe that difficult work is something they can approach rather than avoid.

Paraeducators help make that possible.

You are not there to reduce rigor. You are there to protect it. You are guardians of rigor. Not reducers of it.

Emotional Considerations

Students receiving accommodations may experience complicated emotions during learning. Support can make school more accessible, but it can also make differences visible. Some students may feel different from their peers. Others may feel embarrassed when support is noticeable. Some experience frustration when the work remains challenging even with help. Over time, repeated adult assistance can also create a sense of overdependence if the student begins to believe they cannot work without someone beside them.

Because of this, dignity becomes an essential part of effective support. The goal is not simply helping a student complete the task. The goal is protecting the student's identity as a capable learner while providing the access they need.

Dignity-first support often includes small but powerful choices: A neutral and respectful tone. Private check-ins instead of public correction. Avoiding visible adult takeover of the task. Reinforcing the student's capability. Naming the strategies the student is using rather than emphasizing limitations. These choices communicate an important message: the student remains the thinker.

Language also shapes how students interpret support. The words adults use can either reinforce capability or unintentionally highlight limitation.

For example, saying: "You need help." can quietly signal that the student is less capable. A different phrasing communicates something stronger: "Let's use the strategy that supports you."

The focus shifts from deficiency to skill. The support becomes a tool rather than a label.

Over time, these small language choices matter. Students hear them repeatedly during moments when learning feels difficult. Those moments accumulate, and the messages embedded within them begin to shape how students see themselves. Identity is shaped through repetition.

IMPLEMENTATION

One IEP Alignment Check

Small adjustments in support often begin with careful attention to one student at a time. Rather than attempting to change every routine at once, focus on a single learner and examine how your current support aligns with their IEP goals.

This week, select one student you regularly support and take a few minutes to reflect on how assistance is being provided during learning tasks.

Begin with these steps:

1. Review the student's IEP goal so the purpose of the support is clear.
2. Identify one support you may be using too frequently or for too long.
3. Plan one small way to fade that support while keeping the student successful.
4. Pay attention to how the student attempts the task when support steps back slightly.
5. Communicate with the teacher about what you noticed and discuss whether the adjustment aligns with the instructional plan.

As you observe the student during the week, look for small changes: Did the student attempt the work more regularly? Did confidence shift, even in small ways?

When support fades gradually, progress may feel slower at first. Students sometimes hesitate when a familiar form of help changes.

This hesitation is often a sign that the student is beginning to rely on their own thinking rather than waiting for immediate assistance. With time and consistent encouragement, those moments of uncertainty give way to greater confidence and independence.

Closing Reflection

Special education support carries enormous influence in the classroom. The way support is provided shapes how students experience learning, challenge, and themselves as learners.

It can build skills, strengthen confidence, increase access to learning, and preserve student dignity. But support can also move in a different direction when it is not aligned carefully with learning goals. It can also lower expectations unintentionally, create patterns of dependence, and reduce student ownership of thinking.

The difference rarely comes from a single decision. It develops through repeated interactions during everyday classroom moments. When support consistently protects thinking, students experience themselves as capable. When support replaces thinking, students may begin to believe they cannot succeed without someone beside them.

This is why the role of the paraeducator is so important in inclusive classrooms. Paras are not simply helpers in inclusion. You play a much deeper role in how students experience school.

You are access architects who help students reach the same learning goals through different pathways. You are dignity protectors who ensure support never exposes or diminishes the student. You are independence designers who help students grow stronger thinkers over time

When accommodations remain aligned with rigorous expectations, students build real skill rather than simply completing tasks. When independence is planned intentionally, support becomes a bridge rather than a replacement. And when dignity remains intact during moments of difficulty, something powerful happens. Students begin to see themselves differently. They begin to believe they belong in the work. They begin to believe they can grow.

In that moment, inclusion becomes more than access. It becomes empowerment.

PART V — PROTECTING YOURSELF

Chapter 9: Authority Without Escalation

Paraeducators live in a complicated professional space. You are not the classroom teacher. But you are an adult in the room. You are not writing referrals. But you are often the first person responding when behavior shifts. You are not setting school policy. But you are helping enforce expectations. You are not leading the lesson. But you are managing the learning environment beside the desk.

This space between roles can feel unclear at times. Students may test boundaries. Directions may be questioned. Moments of frustration can appear suddenly during independent work, transitions, or group activities. And when students push back, that ambiguity can feel destabilizing.

Many support professionals worry about how much authority they actually hold. They may hesitate to redirect behavior. They may worry about overstepping the teacher's role. Or they may feel pressure to respond quickly when a situation becomes tense. But effective authority in classrooms does not come from titles alone. It comes from presence.

Students notice tone. They notice consistency. They notice how adults respond when frustration appears. Authority grows when students experience adults as calm, predictable, and steady in moments of difficulty.

This chapter focuses on how paraeducators can maintain that steadiness. Not through dominance. Not through power struggles. Not through escalating conflict. But through quiet, consistent authority.

Authority does not require raising your voice. It does not require winning arguments. And it rarely depends on who has the final word. Instead, authority grows when adults remain regulated while students are not. When expectations stay clear even when emotions rise. When boundaries are communicated calmly rather than defensively.

Escalation may feel powerful in the moment, but it rarely produces influence. Influence grows through stability.

The strategies in this chapter will help you:

- Maintain authority without entering power struggles
- Respond to challenging behavior without escalating the situation
- Protect your own emotional energy during difficult moments
- Reinforce classroom expectations while maintaining student dignity
- Stay aligned with the teacher while still responding effectively in the moment

Because protecting yourself matters.

Paraeducators encounter challenging situations throughout the day. Students may test limits, react emotionally, or resist directions when the work feels difficult. When these moments occur repeatedly, they can become exhausting.

Sustainable support work requires tools for maintaining calm authority without absorbing the emotional intensity of every interaction.

When authority is steady, classrooms remain safer for everyone. When escalation decreases, learning continues. And when adults protect their own stability, they are better able to support the students who rely on them most.

Authority vs. Influence

Authority and influence are closely connected in classrooms, but they are not the same thing. Authority comes from the role you hold in the learning environment. It signals that you are responsible for maintaining expectations, protecting safety, and supporting the structure of the classroom. Authority says: "I am the adult in this space." Influence develops differently. Influence grows from the relationships students build with adults over time. It is shaped by trust, consistency, and the way adults respond during difficult moments. Influence says: "You trust my guidance."

Paraeducators benefit from both forms of leadership. Authority establishes the boundaries that keep the classroom stable. Influence helps students accept those boundaries without constant conflict. But influence is fragile.

When interactions escalate, influence often disappears first. Students may comply temporarily when voices rise or tension increases, but long-term trust weakens when authority turns into confrontation. This is why escalation can be so costly in classroom interactions. When pushback happens, students are often testing several things at once. They may be watching to see: Where the boundaries actually exist. Whether expectations remain consistent. How adults communicate under pressure. Whether the adult maintains emotional control. These tests are not always intentional. Sometimes they appear when students feel frustrated, embarrassed, or overwhelmed. Other times they arise simply because students are learning how authority works in social environments. In these moments, the adult response matters.

If the adult reacts emotionally, leverage quickly disappears. If the interaction escalates publicly, the situation can shift into a power struggle where the focus moves away from learning and toward winning the exchange. But when the adult remains steady, something different happens. The boundary remains clear. The tone stays calm. The interaction slows down instead of intensifying. In those moments, authority remains intact. Authority does not need to be loud. Authority does not need to be dramatic. Authority is calm.

When Authority is Tested

Paraeducators often encounter moments when students push back. These moments can feel uncomfortable because the paraeducator role exists in a unique space within the classroom. You are not the classroom teacher, yet you are still responsible for maintaining expectations. You may not assign consequences, but you are still guiding behavior and supporting the learning environment.

Pushback often appears in small ways at first. A student may roll their eyes when redirected or ignore a direction and continue what they are doing. Sometimes the response becomes verbal. Common tensions include: "You're not the teacher." "I don't have to listen to you." Eye-rolling, sarcasm, ignoring directions, or public defiance in front of peers.

These moments can feel personal, especially when they occur publicly or repeatedly. But most of the time, the student is not challenging you as a person. They are testing the boundaries of authority in the classroom. Students may be wondering: Who holds authority here? What happens if I resist? Will expectations remain consistent? How will this adult respond under pressure?

How adults respond in these moments shapes what happens next. Authority rarely comes from volume, argument, or escalation. Instead, authority grows through alignment and consistency. When your expectations match the teacher's expectations and your tone remains calm, students see that pushback does not destabilize the adult in the room.

The goal in these moments is not to win the exchange. Power struggles shift attention away from learning and toward confrontation. Your role is different. You are there to stabilize the environment so learning can continue. You are there to maintain expectations without escalating the moment.

You are not there to win. You are there to stabilize.

Calm Response Script

When pushback occurs, the goal is not to out-argue the student. The goal is to remain steady.

Students often push verbally when they feel frustrated, embarrassed, or when peers are watching. In those moments, emotional reactions from adults can unintentionally escalate the situation. A raised voice, a defensive tone, or a long explanation can turn a brief moment of resistance into a power struggle.

Instead, rely on neutral, repeatable language. Short statements communicate that the expectation remains stable without inviting debate. They also help you stay regulated when a situation begins to feel tense.

For example: Student: "You're not my teacher." You: "I'm part of the adult team in this room."

Student: "I'm not doing it." You: "The expectation is still the same."

Student: "You can't tell me what to do." You: "I'm here to support this class running smoothly."

Student: "This is stupid." You: "You don't have to like it. You do need to try."

Notice the structure of these responses. They are brief. They restate the expectation. And they do not argue with the student's emotion.

Students are allowed to feel frustrated. They are not allowed to remove the expectation. The adult response simply holds the boundary.

In many situations, the tone of your voice matters more than the exact words you use. Students are very sensitive to changes in adult tone and body language. If the adult appears defensive or angry, students often respond with more resistance. Steady tone communicates control.

Helpful signals include:

- A slow, even speaking pace
- A neutral facial expression
- Minimum increase in volume
- Relaxed posture and steady eye contact

These signals communicate something important: the adult in the room is regulated.

Short statements also serve a practical purpose. They reduce the chance that the interaction turns into a debate.

When adults explain too much during moments of pushback, students sometimes treat the explanation as an invitation to argue every point.

A simple rule can help in these moments: Short statements signal stability. Long explanations invite debate.

You are not trying to win the conversation. You are maintaining the structure of the classroom. And when the adult remains calm and consistent, most moments of pushback lose momentum quickly.

De-Escalation Ladder

Escalation in classrooms rarely happens all at once. Most conflicts develop gradually, moving through predictable stages. Understanding these stages helps adults respond earlier, before the situation intensifies. When you recognize the pattern, you can interrupt the escalation instead of unintentionally climbing the ladder with the student.

Think of conflict as a ladder. Each step increases emotional intensity. Your goal is to stabilize the situation before it rises higher. These steps move from student frustration, verbal pushback, public challenge, emotional escalation and safety risk. How you respond at each step helps or hinders student escalation.

Step 1: Student Frustration

The first signals are often subtle. A student's tone changes. Body language shifts. They may sigh, tap a pencil, slump in their chair, or stop working. These behaviors often signal frustration, confusion, or fatigue. At this stage, the student is still regulated enough to respond to gentle support.

Your move: Calm redirection. A quiet check-in or brief prompt often resets the moment.

Examples include: "Let's refocus on the next step." "Show me what you've tried so far." "Start with the first step."

Early intervention here prevents many conflicts from developing further.

Step 2: Verbal Pushback

If frustration continues, students may begin testing boundaries verbally. Sarcasm, refusal, or dismissive comments can appear.

You might hear statements such as: "I'm not doing this." "This is stupid." "You're not my teacher."

At this stage, the student is beginning to challenge expectations.

Your move: Neutral boundary statement.

Respond briefly and calmly. Avoid arguing or explaining excessively.

Examples include: "The expectation is still the same." "You don't have to like it. You do need to try." "Let's get started with the first step."

The goal is to restate the boundary without escalating the tone.

Step 3: Public Challenge

Sometimes pushback becomes more visible. The student may raise their voice, make comments in front of peers, or attempt to turn the interaction into a public conflict. When peers become an audience, students may feel pressure to continue the challenge.

Your move: Lower your voice. Remove the audience. Use a short directive.

Examples include: "Let's talk for a moment over here." "Take a breath and start with the first step." "We'll handle this quietly."

Reducing the audience lowers the social pressure that often fuels escalation.

Step 4: Emotional Escalation

At this stage, emotions increase rapidly. Voices may rise. The student may argue, become visibly upset, or struggle to regulate their reaction. This is no longer a moment for extended discussion.

Your move: Reduce words. Increase physical space. Offer a limited choice within a clear boundary.

Examples include: "You can start the work now, or take a short reset and begin in two minutes." "Let's pause and regroup."

Short statements help prevent the situation from intensifying further.

Step 5: Safety Risk

If behavior reaches a point where safety may be affected, the situation moves beyond classroom redirection.

At this stage, follow the teacher's direction and the school's established safety procedures immediately. Paraeducators are not expected to manage serious escalation alone. Protocols exist to ensure the safety of everyone in the room.

The most effective interventions happen early.

When adults notice frustration and respond calmly, many conflicts stop before they grow. When responses remain steady and predictable, students often regulate more quickly.

Your goal is not to win the interaction. Your goal is to interrupt escalation before it climbs higher.

What Escalates Conflict

Moments of conflict in classrooms rarely begin at their most intense point. Escalation usually grows in small steps – a correction, a tone shift, a public comment, a moment of embarrassment. What begins as a minor interaction can quickly become a power struggle if the adult unintentionally adds pressure rather than stability.

Paraeducators and teachers often step into these moments with the best intentions: to restore order, protect learning, or correct behavior quickly. But certain responses, even when well-intended, can intensify the situation instead of calming it.

Understanding what escalates conflict helps you choose responses that protect dignity, reduce emotional intensity, and keep the situation from climbing further up the escalation ladder.

To avoid escalating conflict, avoid:

- **Public correction delivered with a sharp tone.** When correction happens in front of peers, students often experience the moment as exposure rather than guidance. What may feel like simple redirection to the adult can feel like humiliation to the student. When dignity feels threatened, students shift from learning mode to self-protection.
- **Sarcasm.** While sarcasm may appear lighthearted among adults, for students it often communicates ridicule or dismissal. Even subtle sarcasm can signal that the adult is not taking the student seriously.
- **Matching the student's volume or intensity.** When an adult raises their voice in response to a student's raised voice, the interaction quickly becomes a contest of dominance rather than a moment of guidance. Escalation often follows a predictable pattern: the louder one party becomes, the louder the other responds.

- **Power-based language.** Statements such as "Because I said so," "You will do this now," or "Don't test me" emphasize control rather than collaboration. When power becomes the focus, students often shift from cooperation to resistance.
- **Long lectures during conflict.** Extended explanations rarely help in the middle of escalation. When students are emotionally activated, their ability to process complex language drops dramatically.
- **Arguing logic during emotional escalation.** Attempts to debate fairness, explain consequences, or prove a point often fail in these moments. The student may not disagree because the logic is flawed, but because their emotional system is already activated.

Once emotional regulation drops, logic does not land. Neuroscience research confirms what you may observe in classrooms. Heightened emotion does not share space with reasoning. When the brain perceives threat or intense stress, activity in the amygdala increases while access to the prefrontal cortex – the region responsible for reasoning, problem solving, and impulse control – becomes temporarily reduced (Siegel, 2012). In these moments, students are not choosing to ignore logic; their brains are prioritizing emotional safety over analytical thinking. This is why attempts to "talk a student through it" during escalation often fail.

Stability must precede reasoning. Calm tone, brief language, and physical space help signal safety to the nervous system. Once emotional intensity decreases, the brain becomes more available for reflection, accountability, and problem solving. At that point, conversation and reasoning can be effective. The goal in moments of escalation is not to win the interaction. It is to restore regulation so learning and problem solving can begin again.

When adults respond with steadiness rather than intensity, escalation often stops earlier in the ladder. The classroom remains safer, relationships remain intact, and students are more able to return to learning once the moment passes.

Professional Boundary Guide

Paraeducators often work in complex spaces. Clear professional boundaries help protect both you and the classroom environment. They reduce confusion for students, maintain alignment with the teacher's instructional plan, and ensure that responsibility is shared appropriately among adults in the room. When boundaries are clear, support becomes stronger and more consistent.

Academic Boundaries. Paraeducators play an important role in helping students access instruction. Clear academic boundaries help maintain alignment with the teacher's plan and ensure that students experience consistent expectations throughout the classroom.

- Refer grading questions to the teacher. Students may ask whether an assignment will count or whether their grade can be adjusted. Referring students to the teacher protects transparency and prevents misunderstandings.
- Collaborate with the teacher before adjusting assignments. If a student is struggling significantly, share that observation with the teacher so appropriate adjustments can be considered. Maintaining communication ensures that any modifications support the learning goals of the lesson and student.
- Clarify directions privately when needed. If instructions appear unclear, check with the teacher privately rather than revising directions in front of students. This helps maintain instructional consistency and protects classroom clarity.

Behavioral Boundaries. Students test limits when expectations feel uncertain. Consistent behavioral boundaries help create a structured environment where students understand expectations and feel safe participating in the learning process.

- Reinforce classroom norms consistently. Students benefit when expectations remain stable across adults in the room as this creates feelings of predictability and fairness.
- Use calm redirection instead of debate. When students become frustrated or resistant, brief and steady redirection is often more effective than extended discussion.
- Involve the teacher in high-level decisions. Major behavioral decisions – such as discipline referrals or significant consequences – belong to the teacher. When situations move beyond routine redirection, involving the teacher helps ensure decisions remain consistent.
- Share observations when patterns emerge. If certain behaviors appear repeatedly, brief documentation or communication with the teacher can help identify patterns and support appropriate responses.

Emotional Boundaries. Working closely with students means experiencing a wide range of emotions throughout the day. Healthy emotional boundaries help adults remain steady, professional, and supportive even during challenging moments.

- Recognize that student frustration is not personal. Students can say hurtful things when they feel overwhelmed or embarrassed. These comments often reflect the student's emotional state rather than the adult supporting them.
- Respond to pushback with calm perspective. Resistance can appear when students feel confused or uncertain. Viewing pushback as information rather than rejection helps maintain a steady response.
- Maintain a steady adult presence. Students take emotional cues from adults. A calm tone, relaxed posture, and measured response can help de-escalate tension and restore stability in the classroom.

Clear boundaries reduce confusion, protect professional roles, and create a more stable learning environment. When students know where decisions are made and how expectations are reinforced, the classroom becomes calmer for everyone.

Boundary Clarifying Language

Students sometimes ask questions that fall outside a paraeducator's role. In those moments, clear and calm language helps maintain alignment with the teacher while protecting the structure of the classroom. Boundary statements do not need to be long or complicated. In fact, the most effective responses are brief, neutral, and consistent.

When boundaries are communicated calmly, students quickly learn where decisions are made and what the expectations are. The goal is not to shut down conversation, but to redirect it to the appropriate place while maintaining respect and clarity. Use phrases like:

"That's a teacher decision." This reminds the student that certain academic or behavioral decisions belong to the teacher.

"Let's bring that to your teacher." This keeps the conversation collaborative while guiding the student toward the right person.

"I'm here to support the plan already in place." This reinforces that your role is to help students work within the expectations the teacher has established.

"We can talk about that calmly." This helps slow emotional escalation and signals that respectful communication matters.

"The expectation remains the same." This maintains consistency when students attempt to negotiate or change expectations in the moment.

Short. Calm. Consistent. Boundary language works best when it is delivered without frustration or explanation. When adults repeat the same calm responses over time, students begin to understand that expectations are stable and predictable.

Elementary Example

A student says loudly: “You’re not the boss of me!”

The comment is sharp and public. Other students turn their heads. A few pause their work to watch.

Moments like this can quickly become a power struggle if the adult responds with equal intensity. When adults match the student’s volume or challenge the statement directly, the interaction can shift from guidance to argument.

Instead of escalating, you respond calmly: “I’m part of the adult team here” or “I’m here to help keep our classroom safe and focused” or “You don’t have to like the direction, you do need to follow it.”

Your tone stays steady. Your voice lowers rather than rises.

Then you add: “Let’s step aside for a minute” or “We’ll solve this together, but we need calm voices” or We’re going to pause for a minute and then rejoin the group.”

You move a few steps away from the group. The audience disappears. The student is no longer performing for peers.

Your goal is not to win the exchange. Your goal is to restore calm and reestablish structure.

You reduce the audience. You reinforce adult authority without arguing. And once the moment settles, the student is more able to return to the work.

Middle School Example

Student rolls eyes and says: "Whatever."

Middle school students often test tone as much as rules. Sarcasm, eye-rolling, and dismissive comments can invite adults into a verbal contest. If the adult corrects the tone directly, the exchange can stretch into a longer confrontation.

Instead of confronting the tone, you respond simply: "The direction still stands" or "You can choose how you respond, the expectation doesn't change" or "Let me know when you're ready to begin" or "Start with the first step", or "We're continuing the task."

Your voice remains neutral. You do not debate the sarcasm. You do not ask the student to explain the attitude. You do not extend the moment longer than necessary.

Then you step away.

Stepping away communicates confidence. It signals that the expectation remains in place without requiring an argument to reinforce it.

You do not chase sarcasm. You maintain the expectation. And the student has space to re-engage without losing face in front of peers.

High School Example

Student refuses to begin work. “I’m not doing this.”

In high school classrooms, refusal sometimes appears as a direct challenge. Students may test whether the adult will debate the assignment or attempt to persuade them in the moment. Extended debate often increases resistance.

Instead, you respond calmly: “You don’t have to agree with it. You do need to attempt” or “Start with one part” or “You don’t have to agree with the assignment to start it” or “Let’s begin with the first question.”

This statement separates opinion from expectation. The student is not required to like the assignment, but the expectation to try remains.

If the refusal continues, you avoid turning the moment into a prolonged argument. You say: “I’ll give you a minute. I’ll check back” or “You don’t have to finish it all right now, just start” or “Take a minute, then begin” or “Show me the first step when I come back.”

Then you step away.

This short pause allows the student time to reconsider without feeling cornered. It protects dignity while keeping the expectation intact.

You avoid debate. You preserve dignity. You keep control of the tone in the room.

Calm authority often resolves situations more effectively than confrontation. When adults remain steady, expectations stay clear, and many students eventually choose to re-engage.

The Emotional Reality for Paras

Working closely with students means you experience emotions. You sit beside students when they are confused, frustrated, tired, or overwhelmed. You see the hesitation before others notice it. You hear the tone shift. You feel the tension in moments when learning becomes difficult. Because of this proximity, student reactions can feel directed at you – even when they are not. Pushback can feel personal.

You are physically and relationally near students. Their words land closer. You have invested in them. You have spent time supporting, encouraging, and helping them succeed. When resistance appears, it can feel like that effort is being rejected. You feel disrespected. Tone, language, or body language can trigger a strong emotional response, especially in public moments.

These reactions are real. And they are human. But behavior is often about control – not about you. When students feel uncertain, exposed, or overwhelmed, they may try to regain control in the only ways they know: refusal, sarcasm, avoidance, or defiance. These responses are not usually personal attacks. They are attempts to manage discomfort.

Understanding this shifts your response. Instead of reacting to the words, you respond to what is underneath the behavior. Stay adult. You are the steady presence in the interaction. You do not need to match the student's tone or energy. Stay regulated. Your calm nervous system helps signal safety. When you remain steady, escalation is less likely to continue. Stay brief. Long explanations and emotional reactions often add intensity. Short, neutral language keeps the moment contained. The calmer adult holds authority.

Authority in these moments comes from consistency, predictability, and emotional steadiness. Students often test limits to see what will happen next. When the response is calm, clear, and consistent, the testing usually decreases over time. You are not responsible for every reaction a student has. You are responsible for how you respond. And that response shapes whether the moment escalates – or settles.

Protecting Your Energy

Escalation drains paraeducators quickly. These moments are not just instructional. They are emotional. They require energy, attention, and regulation – often all at once. When escalation becomes frequent, the impact builds. Not all at once, but over time. If you:

- Argue often, you keep the focus on conflict rather than learning. Arguments require constant engagement and can leave you feeling depleted.
- Match emotion (frustration with frustration or intensity with intensity), the situation often grows instead of settles. Matching emotion increases the demand on your own regulation.
- Over-explain you expel energy while often adding confusion or resistance when the student is not ready to process them.
- Personalize behavior you can make each interaction feel heavier. Over time, this can create emotional fatigue and reduce your capacity to stay steady.

You will burn out. Not because you are doing the work incorrectly, but because you are carrying more than the role requires. Escalation pulls you into every moment. It asks you to respond emotionally, repeatedly, and without pause.

Authority without escalation protects your nervous system. When you respond with calm, brief, and consistent language, you reduce the emotional load on yourself. You are not entering every conflict. You are containing it.

You do not need to win the moment. You need to hold the structure. It also protects student dignity.

When adults avoid escalation, students are given space to recover without being exposed or pushed further. The interaction stays smaller. The relationship remains intact. Calm authority is protective – for you and for the student.

IMPLEMENTATION

Change in classrooms rarely comes from large shifts. It begins with small, repeated actions. Moments of escalation happen quickly. Your response becomes the pattern students learn. This week, focus on one intentional adjustment.

1. **Choose one Calm Response Script.** Select a short, neutral phrase you can use consistently (e.g., "The expectation remains the same" or "I'll check back in a minute"). Keeping it simple makes it easier to use in real time.

2. **Practice using it neutrally.** Say the phrase with a steady tone, relaxed posture, and minimal added language. The goal is not to persuade or correct emotion. The goal is to communicate expectation clearly and calmly.

3. **Notice when escalation begins.** Pay attention to early signals – tone shifts, body language changes, increased volume, or resistance. Escalation often starts before it becomes obvious.

4. **Interrupt early using the ladder.** Use your calm response before the situation grows. Early intervention requires less energy and is more likely to keep the interaction from escalating further.

5. **Reflect: Did my tone stay steady?** After the moment, take a brief pause. Notice not just what you said, but how you said it.

Ask yourself: Did I escalate — or stabilize? Growth in this work is not about perfection. It is about awareness.

Each time you respond with calm, consistent language, you strengthen your ability to hold steady in difficult moments. Over time, those small responses reduce escalation, protect your energy, and create a more predictable environment for students.

Consistency builds confidence – for you and for them.

Closing Reflection

Paraeducators do not need louder voices. They need ones. In moments of tension, it can feel natural to increase volume, add urgency, or assert control more forcefully. But in classrooms, volume rarely builds authority. It often invites resistance. What students respond to over time is not intensity or authority – it is consistency.

Students watch closely. They notice how adults respond when situations become difficult. They notice whether expectations change, whether tone shifts, and whether boundaries hold. When responses are steady and predictable, students begin to trust the structure of the classroom.

Influence grows when students feel:

- The boundary is firm. Expectations do not shift based on emotion, negotiation, or pressure.
- The adult is calm. Tone, posture, and language remain steady, even when the student is not.
- The expectation is stable. Students know what will happen next because the response is consistent.

These conditions create safety. And safety allows students to re-engage. You do not have to win the moment. You need to hold it. Holding the moment means resisting the urge to argue, explain excessively, or match emotion. It means staying present, clear, and grounded while the situation settles.

And when you hold it without escalating, you protect yourself, you protect dignity, and you protect the learning environment.

Calm authority does not draw attention to itself. But over time, it shapes everything.

Chapter 10: Sustainable Support Work

Paraeducators are often the emotional shock absorbers of the classroom.

You sit beside frustration. You absorb anxiety. You witness meltdowns. You regulate dysregulation. You comfort quietly. You redirect gently. You hold steady when others cannot. These moments are often unseen. They happen beside the desk, in the hallway, during transitions, in the spaces between instruction. While the lesson continues, you are managing the emotional undercurrent that makes learning possible.

And then you go home. Often exhausted. Not just physically. Emotionally.

The work asks more than it appears to from the outside. It requires attention, patience, and constant regulation. It asks you to remain calm when others are not. It asks you to respond thoughtfully in moments that happen quickly. Over time, that steady output of energy can accumulate.

Without intention, the work can begin to follow you. You may replay conversations. You may carry student stories. You may feel responsible for outcomes beyond your control. Sustainable support work requires more than skill. It requires boundaries around what you carry.

This chapter focuses on how to remain effective without becoming overwhelmed – how to support students while also protecting your own energy, clarity, and well-being.

You do not have to carry everything to be effective. In fact, the work becomes stronger when you do not.

Sustainability is not stepping back from the work. It is learning how to stay in it – without losing yourself.

The Invisible Weight: Emotional Labor

Much of the work paraeducators do is not written into a schedule. It is not listed in lesson plans. But it is present in nearly every interaction. This is emotional labor.

Emotional labor is the effort required to stay calm when others are nor, respond kindly when disrespected, regulate your tone when pressure increases, hold space for distress, encourage when progress is slow, and protect dignity during struggle. Paraeducators perform emotional labor constantly. Often it is invisible work. But it consumes real energy.

Research on emotional labor confirms what educators experience daily: managing emotions as part of a professional role requires sustained effort and can lead to emotional exhaustion over time if not supported (Hochschild, 1983). The more consistently individuals must regulate their emotions in response to others, the greater the potential impact on well-being.

If unmanaged, this can lead to irritability where small frustrations begin to feel heavier and harder to manage, compassion fatigue where the ability to respond with patience and care beings to decrease, cynicism, emotional numbness, or burnout.

These outcomes are not a sign of weakness. They signal that energy is being given without being replenished. You cannot sustainably give what you do not replenish.

Sustainable support work requires noticing not only what you give to students, but also what you need to restore in yourself.

Secondary Trauma Is Real

Paraeducators frequently work with students who have experienced trauma, face instability, struggle with regulation, and carry stress from outside of school into the classroom.

You are often the adult who hears these stories. You are present during moments of distress. You sit beside students when emotions rise. Over time, this exposure can accumulate. Listening to repeated distress stories, witnessing dysregulation, or absorbing intense emotional energy can lead to secondary trauma – the emotional impact of supporting others through difficult experiences.

Research on secondary traumatic stress in helping professions shows that individuals who work closely with those experiencing trauma can begin to feel the effects themselves, even when they have not directly experienced the events (Figley, 1995). The impact is not always immediate, but it builds through repeated exposure.

You may notice you carry student stories home, replaying them in your mind after the school day ends. You may think about a specific child constantly, have difficulty sleeping, experience intense emotions, or feel responsible for outcomes beyond your control.

These responses are a sign that you are paying attention. That you care, and caring deeply is not weakness. It is part of what makes this work meaningful. But carrying everything is unsustainable.

Sustainable support requires recognizing what belongs to you – and what does not. You can care, support, and show up consistently without taking on the full weight of every situation.

Holding that boundary protects your ability to keep doing the work well.

Energy Mapping

Not all parts of your job drain you equally. Some interactions restore you. Some deplete you. The difference is not always obvious in the moment. Two situations may look similar on the surface – both require support, attention, and patience – but they leave you feeling very different afterward.

Sustainable support requires awareness. Without awareness, it is easy to move from one moment to the next without noticing what is building. Over time, that accumulation can feel like constant exhaustion, even when the source is not clear. Energy mapping helps you notice patterns. Ask yourself:

- When do I feel most energized during the day? Notice the moments when you feel focused, present, and steady.
- When do I feel most drained? Pay attention to the times when your energy drops quickly.
- Which students require the most emotional regulation from me? Some interactions ask more of your patience, attention, and emotional control.
- Which tasks exhaust me quickly? Certain routines may consistently require more effort, even if they appear simple.
- Which tasks feel manageable? Identifying what feels sustainable helps create balance.

As you reflect, patterns begin to emerge. You may notice that certain times of day are more draining. You may notice that specific types of interactions require more recovery. You may notice that transitions, behavior support, or academic frustration affect your energy differently. Without awareness, exhaustion feels random. It feels like the entire day is equally heavy. With awareness, boundaries become possible. You can prepare for the moments that require more energy. You can pace yourself across the day. You can recognize when you need to pause, reset, or step back briefly.

Energy mapping does not remove the demands of the job. It helps you navigate them with intention. And that awareness is what makes the work sustainable.

Knowing What Is Yours to Carry

Paraeducators often overextend emotionally. The work invites it. You are close to students. You see what others may not see. You care. Over time, that care can begin to feel like responsibility for more than your role was designed to hold. You may feel:

- Responsible for fixing home situations. You hear pieces of a student's life and wish you could change what they return to.
- Responsible for emotional healing. You want the student to feel better, calmer, more secure – and you feel the weight when they do not.
- Responsible for academic outcomes. When a student struggles, it can feel like you should be able to make it click.
- Responsible for behavioral change. You may feel pressure to stop patterns immediately or correct behavior completely.
- Responsible for student happiness. You want students to feel successful and positive in the classroom.

These responses come from care. But they can quietly expand the role beyond what is sustainable.

Here is a grounding truth: You are responsible for your role. You are not responsible for fixing every part of a student's life.

Your role is powerful – and specific. You can support, stabilize, encourage, and protect dignity. These actions matter deeply. They are enough.

You cannot repair every trauma, control every choice, and guarantee every outcome. Holding this boundary does not reduce your impact. It protects it.

When you carry only what is yours, you remain steady. You remain present. You remain able to continue the work. Knowing what is yours to carry protects your longevity. And longevity is what allows your support to matter over time.

Daily Decompression Reflection

The school day does not always end when the bell rings. Moments linger. A conversation replays. A difficult interaction follows you into the evening. Without intention, the day continues in your mind and body.

Daily decompression creates a clear ending. It helps you separate what happened at school from what you carry forward. At the end of each day, take 5 minutes. Pause before moving into the next part of your day. Ask yourself:

1. What went well today? Notice even small moments – a student attempting, a calm response, a brief connection. This helps anchor your attention in what is working.
2. What drained me? Identify where your energy was used most. Naming it brings clarity.
3. Which student interaction is lingering? Pay attention to what is still on your mind. Lingering moments often signal emotional weight.
4. Is that something I can act on tomorrow? If yes, hold it as a next step. If no, it may not belong to you tonight.
5. If not, can I release it tonight? Releasing is a choice. It may not feel immediate, but naming the boundary begins the process.

Then write one sentence: "Today I supported by ______." This centers your role and reinforces what you contributed.

Then write one boundary statement: "What is not mine to carry tonight is ______." Be specific. Name the student, the moment, or the concern.

This is not avoidance. It is intentional separation. Reread what your wrote. Take a breath. Stand up. Signal to yourself that the workday is complete. Do not take the classroom home in your nervous system. The work will continue tomorrow. You will return steadier if you allow yourself to step away today.

Emotional Containment Checklist

The end of the day is not just about leaving the building. It is about leaving the emotional weight behind. Some moments stay with you longer than others. A tone. A look. A difficult interaction. Without awareness, those moments follow you. They replay. Emotional containment helps you notice what you are carrying – and can help you decide what to release.

Before leaving work, ask:

- Did I internalize student frustration?
- Did I interpret a student's reaction as something about me?
- Am I replaying a conflict?
- Did I absorb distress without processing it?
- Did I forget to regulate myself today?
- Did I continue working without pause, even when I needed one?

If yes to multiple, pause. Do not rush past the signal. Your body is asking for a reset. Take three slow breaths. Inhale slowly. Exhale fully. Let your shoulders drop. Then remind yourself: "I can care without carrying everything."

You can be present, supportive, and consistent – without absorbing every emotion you encounter. Containment is not coldness. It is sustainability.

It allows you to return tomorrow steady, clear, and able to continue the work.

Professional Identity Affirmation

Paras sometimes feel undervalued. Invisible. Replaceable. Much of your work happens quietly. It is not always recognized. It is not always named. But your work matters.

You shape how students experience themselves as learners. You influence whether moments escalate or settle. You help build confidence, stability, and independence – perhaps without attention or acknowledgment. When this work is not named, it can begin to feel unclear. And when identity feels unclear, exhaustion can feel heavier.

Clarity changes that. Creating a professional affirmation helps you define your role in your own words. It reminds you what you bring to the classroom – even on difficult days.

Create a personal affirmation statement. Keep it short. Keep it clear. Keep it aligned with how you want to show up. Examples: "I have a steady adult presence." "I build independence." "I regulate calmly." "I support without rescuing." "I contribute to growth." Your statement does not need to be perfect. It needs to be true.

Read it weekly. Or return to it after a difficult day. Over time, this repetition reinforces your role and purpose.

Identity clarity reduces burnout. When you know who you are professionally, you are less pulled in different directions. You are less likely to take on what is not yours.

The work may still be demanding, and it becomes clearer.

Energy Boundaries in Practice

Energy boundaries are not about caring less. They are about carrying appropriately. In each setting, the work invites you to take on more than your role requires. Boundaries help you stay present without becoming overwhelmed.

Elementary

After a student meltdown, the moment can stay with you. You may replay what happened. You may think about what you could have said differently. You may wish you had prevented it.

Instead, pause and remind yourself: "The building stays here, I do not carry it home." Not: "I should have prevented it." In that moment, you helped de-escalate. You protected safety. You supported the student through a difficult experience. That matters.

You are not responsible for eliminating every emotional response. You are responsible for how you respond when it happens.

Meltdowns are often part of development and regulation. Your role is to support the moment – not control every outcome.

When the moment ends, your role in it ends as well.

Middle School

A student confides about family stress. The moment is quiet. Often private. Sometimes unexpected. The student may not be asking for advice. They may not be asking for solutions. They are asking to be heard. You listen. You stand with them. You might say: “That sounds really hard.” “I’m glad you told me.”

You do not rush to fix. You do not offer solutions you cannot provide. You do not take on the role of counselor. You hold the moment with care.

You may be one of the few adults the student feels safe talking to. That trust matters. Then you follow appropriate steps. You report to the appropriate adult if needed. You communicate what is necessary to ensure the student is supported.

You stay within your role. After that, you release responsibility for solving it.

You do not carry the full story with you. You do not take on outcomes that are beyond your control. You do not feel responsible for fixing what requires a broader system of support.

Support does not equal ownership.

You can care deeply. You can listen attentively. You can respond appropriately. And still, you are not the solution to every problem shared with you. You are part of a team. A system. A network of support.

When you hold that boundary, you remain steady.

And steadiness allows you to continue showing up for students – without becoming overwhelmed by what you cannot control.

High School

A student refuses work repeatedly. The pattern is familiar. The assignment is placed in front of them. Time passes. Nothing begins. You may feel the pull to intervene more strongly. To persuade. To explain again. To convince the student to care.

Over time, repeated refusal can begin to feel personal. As if your support should have made a difference. As if you should be able to change the outcome.

Pause and return to your role. You implement boundaries. You maintain the expectation. You might say: "I'm looking for a start, not perfection" or "You can begin now or when I come back; your choice." Your tone remains calm. Your language remains brief. You do not argue. You do not negotiate.

If the refusal continues, you stay consistent. You document. You communicate with the teacher. You align with the plan already in place.

You do not carry the refusal as a personal failure. Student decisions are shaped by many factors – motivation, past experiences, confidence, external stress, and readiness. These are not all within your control.

Your role is to provide structure, clarity, and steady support. You offer the opportunity. You maintain the expectation. You create space for re-engagement. Whether the student chooses to engage in that moment is not yours to control. Professional detachment protects your mental health.

Detachment does not mean you stop caring. It means you stop carrying responsibility for outcomes that are not yours to hold. You remain present. You remain consistent. You remain steady. And that steadiness is what allows you to continue showing up – even when progress is not immediate.

Sustainable Does Not Mean Detached

Sustainability is not indifference. It is regulated care. It is the ability to remain present, steady, and supportive – without becoming overwhelmed by what you cannot control.

Sometimes boundaries are misunderstood. They can feel like distance. But sustainable support is not about caring less. It is about caring in a way that can continue. You can care, show warmth, and encourage consistency, and at the same time, not overidentifying by taking on a student's emotions as your own, over functioning by doing the work the student is capable of attempting, or holding responsibility for every outcome.

These boundaries are not barriers. They are supports. They allow you to remain steady across the day. They protect your ability to show up again tomorrow. You are part of a system. Not the entire system by yourself. There are teachers, counselors, administrators, families, and community supports. Each plays a role.

When you care with boundaries, your support becomes more consistent, more sustainable, and more effective.

Regulated care does not reduce your impact. It allows your impact to continue over time.

When You Need Support

Paraeducators often hesitate to ask for help. You may not want to interrupt. You may not want to appear unprepared. You may feel that you should already know what to do. Over time, this hesitation can lead to carrying more than your role was designed to hold. But sustainability requires support.

Support is not a sign that something is wrong. It is part of doing the work well. Sustainable support requires:

- Clear communication with teachers. Regular, brief conversations help ensure you understand expectations, priorities, and how to best support instruction.
- Regular check-ins. Touchpoints throughout the week create space to clarify questions, share observations, and stay aligned.
- Professional development. Learning builds confidence. It provides tools, language, and strategies that make daily interactions more effective.
- Emotional validation. The work can be demanding. Having space to acknowledge that reality helps prevent isolation.
- Clear expectations. Knowing your role reduces uncertainty and helps you act with confidence in the moment.
- Breaks. Pauses are not optional. They allow your nervous system to reset so you can remain steady throughout the day.

Support does not always arrive automatically. Sometimes it needs to be requested. Advocate for clarity. Ask questions when expectations are unclear. Advocate for training. Seek opportunities to strengthen your skills. Advocate for alignment. Ensure your support connects to the teacher's plan and classroom systems. Self-advocacy is professional maturity. It reflects awareness of your role, your needs, and the conditions required to do the work sustainably.

When you ask for support, you are not stepping away from your responsibility. You are strengthening your ability to carry it well.

IMPLEMENTATION

Sustainable change does not require a full redesign. It begins with small, intentional shifts. This week, focus on awareness and one adjustment. You are not trying to change everything. You are building one sustainable habit.

1. Complete the Daily Decompression Reflection three times. Choose three days this week to pause at the end of the day. Notice what stayed with you and what can be released. Repetition builds awareness.

2. Identify one recurring drain. Look for a pattern. It may be a time of day, a specific task, or interaction that consistently requires more energy.

3. Adjust one boundary. Choose one small shift. This might be pausing before responding, stepping away instead of staying, or redirecting responsibility back to the appropriate adult.

4. Create one affirmation statement. Write a short sentence that defines how you want to show up. Keep it visible. Return to it during challenging moments.

5. Share one alignment concern with a teacher if needed. If something feels unclear or inconsistent, bring it forward. Alignment reduces confusion and supports more effective practice.

As you move through the week, notice: Does awareness reduce exhaustion? Does simply naming what is happening change how it feels? Does naming boundaries create relief? Do small shifts make the work feel more manageable?

These questions are not about evaluation. They are about noticing. Sustainability begins with permission. Permission to pause. Permission to adjust. Permission to carry only what is yours.

Closing Reflection

Paraeducators give enormous emotional energy. You show up for students in moments that are often difficult, unpredictable, and unseen. You respond with patience when things are not calm. You remain steady when others cannot. But you are not infinite.

Without intention, the work can begin to take more than it gives. Energy can be spent without being restored. Care can become over-carrying. This can be prevented by:

- Boundaries. Knowing what is yours to carry – and what is not.
- Awareness. Noticing when your energy is being used and where it needs to be restored.
- Regulation. Maintaining a steady presence, even in challenging moments.
- Clarity. Understanding your role so you are not pulled in multiple directions.
- Alignment. Working in coordination with the teacher and the broader support system.

These are not extras. They are what make the work sustainable. You are not responsible for carrying the entire weight of the classroom. You are responsible for showing up steadily within your role. That steadiness matters. It builds trust. It creates consistency. It allows students to re-engage.

When you protect your energy, you protect your longevity. And longevity is what allows your work to compound over time. Not just in one moment. Not just in one day. But across months and years. And that is where the impact deepens: Building independence. Protecting dignity. Supporting growth.

For years – not just for one exhausting term.

IMPLEMENTATION

30-Day Plan

Change in classrooms comes from doing a few things more intentionally. This 30-day plan is designed to help paraeducators shift from helping students complete work to helping students become more independent thinkers. Each week builds on the last, creating small, sustainable changes that strengthen both student growth and your role clarity.

Week 1 — Observe support patterns. Begin by noticing. Do not change anything yet. Pay attention to what is already happening. Notice:

- When you step in to help
- How quickly support is given
- Whether students attempt before looking to you
- How often you speak compared to the student

This week is about awareness. Patterns that feel automatic become visible when you slow down and observe. Awareness is the foundation for intentional change.

Week 2 — Reduce over-prompting. Once patterns are visible, begin to adjust. Focus on giving students space before stepping in. Try:

- Pausing before responding
- Waiting for a first attempt
- Reducing how quickly you provide hints
- Stepping away instead of staying beside the student

This may feel uncomfortable at first. Students may look to you more. Silence may feel longer. That is part of the shift. Reducing over-prompting allows students to experience themselves thinking.

Week 3 — Increase question-based scaffolding. Replace answers with questions. Instead of telling, guide thinking. Use prompts such as:

- "What do you think the first step might be?"
- "What have you tried so far?"
- "Can you explain your thinking?"

Questions keep ownership with the student. They position the student as the thinker. Your role becomes support – not solution.

Week 4 — Track independence growth. Now, begin to notice change. Look for:

- Increased student attempts
- Longer periods of independent work
- More student explanations
- Less immediate reliance on adult help

Growth may be small at first. It may not appear in every moment. But patterns will begin to shift. Tracking these changes helps make progress visible – for you and for the student.

This plan is not about perfection. It is about progress.

Small shifts in daily interactions can lead to meaningful changes in how students experience themselves as learners. And over time, those changes build independence – which is the goal of support.

APPENDICES

- Quick Reference Questions
- Prompt Fading Ladder
- Behavior Reframing Chart
- AI Use Quick Decision Chart
- Independence Growth Tracker
- Teacher-Para Communication Template
- PLC Facilitation Guide for Para Teams

The work of supporting students happens in real time. In moments that move quickly. In interactions that require judgment, not scripts.

The tools in this section are designed to support that reality. They are not additional tasks. They are supports for the work you are already doing.

Each appendix provides a practical structure to help you: Maintain clarity in your role. Support student thinking without over-supporting. Respond to behavior with intention. Align with teachers and teams. Protect your own energy and sustainability.

These tools are meant to be used, not just read. Some can be copied and kept nearby. Some can be revisited during reflection. Some can be used in team conversations to build shared understanding.

You do not need to use all of them at once. Start with one. Apply it consistently. Notice what changes. Over time, these small structures help create larger shifts: Less reactivity. More clarity. More consistent support. More student independence.

The goal is not perfection. The goal is alignment – between instructional intention and daily interaction; between your role, your actions, and student needs. These tools help bridge that gap.

Appendices A: Quick Reference Questions

Purpose: Shift from telling → thinking. This tool helps you respond to students in ways that keep ownership of thinking with them rather than transferring it to you.

Use when: A student says, “I don’t know.” A student asks for the answer. A student stops working or waits for help. You feel the urge to explain or tell. You want to build independence through questioning.

How to use: Pause before responding. Choose one question; keep it simple. Wait and allow time or the student to process and respond. Follo up with a second question only if the student remains stuck.

Starting questions to initiate thinking: What do you think the first step is? What is the question asking? Where could you begin?

Thinking questions to deepen reasoning: What have you tried so far? Why did you choose that strategy? Can you explain your thinking?

Redirection questions to move forward: What could you try next? Where might you check your work? What does your example show?

Reflection questions to consolidate learning: What worked well? What would you do differently next time? What did you learn from this?

Reminder: Do not replace thinking. Prompt it. Ask one question. Then wait. Silence is part of the process. It gives thinking space. Over time, consistent questioning helps students rely less on adult answers and more on their own reasoning – which is the goal of support.

Appendices B: Prompt Fading Ladder

Purpose: Move from dependence → independence. This tool helps you adjust your level of support so students do more of the thinking over time.

Use when: A student relies heavily on adult help. You notice yourself giving answers quickly. You want to increase student independence. A student is capable but hesitant to begin. You need a quick guide for how much support to provide.

How to use this ladder:

1. Identify your current level of support. Where are you on the ladder?
2. Provide only what is needed. Avoid moving higher than necessary.
3. Gradually reduce support. As the student becomes more confidence, step down the latter (to a higher number).

Level 1 — Direct Support. Adult provides answer or completes task. Use sparingly. This level limits student thinking if overused.

Level 2 — Step-by-Step Guidance. Adult walks student through process. You might say, "First do this...". Provide structure and keep the adult leading.

Level 3 – Guided Support. Adult provides partial support and allows some student struggle. You might say, "Look at this part..." Shift some responsibility to the student.

Level 4 — Questioning. Adult prompts thinking by asking questions. You might say, "What do you notice?" Ownership of thinking stays with the student.

Level 5 — Independent Attempt. Student works without support. Adult observes and allows the student to problem-solve and create independently.

The goal is to move down the ladder over time. Less direct support → more student independence. When you reduce support intentionally, students begin to rely more on themselves – which is the goal of learning.

Appendices C: Behavior Reframing Chart

Purpose: Shift from reaction → interpretation. This tool helps you pause, look beneath the behavior, and respond to the underlying need rather than reacting only to what is visible.

Use when: A student's behavior feels challenging or repetitive. You are unsure how to respond in the moment. Reactions are becoming emotional or automatic. You want to reduce escalation and increase clarity. You need a quick reset to respond intentionally.

How to use this chart:

1. Notice the behavior. What is happening on the surface.
2. Consider the possible need. What might the student be experiencing underneath?
3. Choose a response. Respond in a way that addressed the need while maintaining expectations.

Behavior	Possible Need	Adult Response
Refusal	Avoidance or overwhelm	Offer structured start
Disruption	Confusion or low confidence	Redirect calmly
Shutdown	Fear or low confidence	Provide small entry point
Defiance	Control or autonomy	Offer bounded choice
Off-task	Fatigue or disengagement	Re-engage with clarity

Remember, behavior is communication. Students often show what they cannot yet say. When you interpret behavior as information, your response becomes more effective. You can respond to the need – not just the action. This shift reduces escalation, protects student dignity, and creates a more stable learning environment.

Appendices D: AI Use Quick Decision Steps

Purpose: Protect student thinking in AI-supported environments. This tool helps you make quick, intentional decisions about when and how AI can support learning without replacing it.

Use this when: A student is using AI tools (e.g., ChatGPT, Grammarly, calculators, etc.). You are unsure whether AI use is helping or replacing thinking. Students are completing work quickly with little evidence of understanding. You want to maintain rigor while allowing appropriate support. You need a quick, in-the-moment decision guide

Step 1: What is the task?

- Learning new content → Limit AI.
- Practicing skills → Guide AI use.
- Checking work → Allow AI support.

Step 2: Has the student attempted?

- No → Require attempt first. Thinking comes before assistance.
- Yes → AI can support reflection. AI can extend or refine thinking.

Step 3: What is AI doing?

- Replacing thinking → Pause
- Supporting thinking → Allow

Step 4: Add Human Layer; ask:

- Explanation – "Can you explain this in your own words?"
- Justification – "Why does this answer make sense?"
- Reflection – "What did you learn from this?"

Remember, AI can assist, and your role is to protect student thinking.

Appendices E: Independence Growth Tracker

Purpose: Make independence visible. This tool helps you notice small shifts in student behavior over time and reinforces a focus on thinking, not just completion.

Use this when: You want to reduce over-supporting. You are working to build student independence. A student relies heavily on adult help. You want to track growth across a week. You need concrete data to share with a teacher or team. You want to make progress visible – even when it feels slow.

Student: ____________

Week: ____________

Day	Attempted First	Needed Prompt	Completed Independently
Mon	☐ Yes ☐ No	☐ Yes ☐ No	☐ Yes ☐ No
Tue	☐ Yes ☐ No	☐ Yes ☐ No	☐ Yes ☐ No
Wed	☐ Yes ☐ No	☐ Yes ☐ No	☐ Yes ☐ No
Thus	☐ Yes ☐ No	☐ Yes ☐ No	☐ Yes ☐ No
Fri	☐ Yes ☐ No	☐ Yes ☐ No	☐ Yes ☐ No

Reflection:

- What improved? Look for small changes – increased attempts, reduced prompting, or longer independent work time.
- What still needs support? Identify where the student continues to rely on adult assistance.

Remember, independence develops gradually. A student attempting first one time is progress. A student needing one less prompt is progress. A student completing part of a task independently is progress. The goal is not perfection. The goal is movement. When you track independence, you begin to see growth that might otherwise go unnoticed. And what you notice, you can intentionally support.

Appendices F: Communication Template

Purpose: Increase alignment and clarity between teachers and paraeducators. Clear communication reduces confusion, prevents over-supporting, and ensures students experience consistent expectations across adults.

Use when: Beginning a new week or unit. Roles or expectations feel unclear. Student support feels inconsistent. You want to better align with the teacher's instructional plan. You are supporting multiple students or groups. You need quick, structured communication (5-10 min).

Weekly Check-In

1. Identify the instructional focus of the week. What is the priority for student learning?(Example: writing structure, math problem-solving, reading comprehension)
2. Identify how your support best fits: Prompting. Monitoring. Behavior support. Small group facilitation. This clarity ensures your support aligns with instruction – not replaces it.
3. Identify what should be avoided. Are there supports that may unintentionally reduce thinking or independence? (Example: over-prompting, giving answers, modifying tasks without direction)
4. Identify which students need your attention. Which students should you prioritize this week? What specific support do they need?
5. Name what questions remain and what needs further clarification.

Remember, the goal is one shared plan. You are not working in parallel – you are working together. When roles are clear and communication is consistent, support becomes more effective, and students experience stability across the classroom.

Appendices G: PLC Facilitation Guide

Build shared practice and reflection with your Professional Learning Team (PLC) of colleagues. This structure supports consistency across adults, strengthens decision-making, and keeps student support front and center.

Use When: You want to improve consistency in how paraeducators support students. You are working to align prompting, behavior support, or independence strategies. A team is experiencing similar challenges across classrooms. You want to shift from reacting individually → solving collaboratively. You need a structured, efficient meeting format (30–45 minutes). Consider this meeting outline:

1. Opening reflection (5 min). What worked well this week? What celebrations do you have to share?
2. Practice focus (10 min). Detail how prompting, behavior support, or increasing independence is working across settings. Describe how this is showing up across settings. Look for patterns, not isolated moments.
3. Share examples (10 min). Discuss what you tried and how that worked. Keep examples specific and brief. Focus on actions and outcomes rather than opinions.
4. Problem-solve (10 min). Identify what was challenging and how you can adjust if the situation returned. Stay solution-oriented; the goal is refinement, not perfection.
5. Next steps (5 min). Name what you commit to trying next. Make it one action. Keep it small and specific. Naming the action creates accountability. It also allows peers to support follow-through.

Teams do best when they have agreed upon norms. Consider norms like staying solution-focused, sharing honestly, presuming good intent, and keeping student dignity central to conversations.

Remember the goal is consistency across adults. Small, repeated conversations lead to meaningful shifts over time.

REFERENCES

Bandura, A. (1997). *Self-efficacy: The exercise of control.* W. H. Freeman.

Bransford, J. D., Brown, A. L., & Cocking, R. R. (2000). *How people learn: Brain, mind, experience, and school* (Expanded ed.). National Academy Press.

Chi, M. T. H. (2009). Active-constructive-interactive: A conceptual framework for differentiating learning activities. *Topics in Cognitive Science, 1*(1), 73–105. https://doi.org/10.1111/j.1756-8765.2008.01005.x

Dweck, C. S., & Diener, C. I. (1980). An analysis of learned helplessness: Continuous changes in performance, strategy, and achievement cognitions following failure. *Journal of Personality and Social Psychology, 36*(5), 451–462.

Figley, C. R. (Ed.). (1995). *Compassion fatigue: Coping with secondary traumatic stress disorder in those who treat the traumatized.* Brunner/Mazel.

Fisher, D., & Frey, N. (2014). *Better learning through structured teaching: A framework for the gradual release of responsibility.* ASCD.

Greene, R. W. (2014). *The explosive child: A new approach for understanding and parenting easily frustrated, chronically inflexible children* (5th ed.). HarperCollins.

Hall, T. E., Meyer, A., & Rose, D. H. (2012). *Universal design for learning in the classroom: Practical applications.* Guilford Press.

Hattie, J. (2012). *Visible learning for teachers: Maximizing impact on learning.* Routledge.

Hochschild, A. R. (1983). *The managed heart: Commercialization of human feeling.* University of California Press.

Individuals with Disabilities Education Act, 20 U.S.C. § 1400 (2004). https://sites.ed.gov/idea/

Kapur, M. (2008). Productive failure. *Cognition and Instruction, 26*(3), 379–424. https://doi.org/10.1080/07370000802212669

Perry, B. D., & Winfrey, O. (2021). *What happened to you? Conversations on trauma, resilience, and healing.* Flatiron Books.

Rosenshine, B. (2012). Principles of instruction: Research-based strategies that all teachers should know. *American Educator, 36*(1), 12–39.

Seligman, M. E. P. (1975). *Helplessness: On depression, development, and death.* W. H. Freeman.

Siegel, D. J. (2012). *The developing mind: How relationships and the brain interact to shape who we are* (2nd ed.). Guilford Press.

Sweller, J., Ayres, P., & Kalyuga, S. (2011). *Cognitive load theory*. Springer.

Tomlinson, C. A. (2014). *The differentiated classroom: Responding to the needs of all learners* (2nd ed.). ASCD.

Made in United States
Orlando, FL
05 November 2026

90231070R00108